Vest Pocket Guide to the National Electrical Code®

VEST POCKET GUIDE TO THE NATIONAL ELECTRICAL CODE®

1996 EDITION

MARVIN J. FISCHER, P.E.

Prentice Hall P T R
Upper Saddle River, NJ 07458

Editorial/production supervision: Rainbow Graphics, Inc.
Manufacturing buyer: Alexis Heydt
Cover Design Director: Jerry Votta
Cover Design: Karen Salzbach
Acquisitions Editor: Bernard Goodwin

Prentice-Hall, Inc.
A Simon & Schuster Company
Upper Saddle River, New Jersey 07458

The publisher offers discounts on this book when ordered in bulk quantities. For more information, contact:

Corporate Sales Department
Prentice Hall PTR
One Lake Street
Upper Saddle River, NJ 07458
Phone: 800–382–3419
FAX: 201–236–7141
E-mail: corpsales@prenhall.com

Printed in the United States of America

10 9 8 7 6 5 4 3 2 1

90000
9 780134 535494
ISBN 0-13-453549-9

Prentice-Hall International (UK) Limited, *London*
Prentice-Hall of Australia Pty. Limited, *Sydney*
Prentice-Hall Canada Inc., *Toronto*
Prentice-Hall Hispanoamericana, S.A., *Mexico*
Prentice-Hall of India Private Limited, *New Delhi*
Prentice-Hall of Japan, Inc., *Tokyo*
Simon & Schuster Asia Pte. Ltd., *Singapore*
Editora Prentice-Hall do Brasil, Ltda., *Rio de Janeiro*

CONTENTS

4 Equipment for General Use 148

5 Special Occupancies 194

LIST OF TABLES

PREFACE

The *Vest Pocket Guide to the National Electrical Code®* is intended for use as a convenient reference guide for architects, engineers, electrical designers, electrical contractors, electricians, manufacturers, inspectors, teachers, students, and all those who use the code in their work. The reader should have a basic knowledge of electricity and the *National Electrical Code®** or be using this book together with other instructional material in the study of electricity, electrical design, and construction. This book is not intended to replace the *National Electrical Code®* but rather be used with it. It is intended to provide the user with a reference book of those topics in the *NEC®** that occur most frequently to the user in a form that can be carried at all times where it might prove inconvenient to carry a copy of the complete *National Electrical Code®*.

The *NEC®* is probably the most referenced and widely distributed single electrical code in use today. It is adapted into law in most jurisdictions in the United States: town, village, city, county, and state. Most federal and state agencies adapt this code in whole or in part or refer to specific pertinent sections and paragraphs. Naturally, there are some exceptions. It is safe to say that if your work or profession in any way touches the electrical field, a thorough knowledge of the *National Electrical Code®* may be quite important, if not an absolute necessity.

Let us examine for a moment who might be using the *NEC®* and for what purpose. The reader will note that it is not uncommon today to find one person falling into more than one category. Those people in the design profession—architects, engineers, and electrical designers—must surely refer to the Code in their everyday work. Not only must the design conform to the Code theoretically, but practically the work must be capable of being installed in conformance with the Code so as to provide a safe installation. It should be noted that the *National Electrical*

* *National Electrical Code®* and *NEC®* are registered trademarks of the National Fire Protection Association, Inc., Quincy, MA 02269.

Code® is not and should not be used as a design specification. This is noted in paragraph 90–1(c) of the Code and will be discussed later. It should be used as a guide for a minimum provision of an essentially safe installation. Adequate space must be provided, and proper wire size and insulation must be used for specific purposes. These are just a few of the many topics of concern for design professionals. All are aimed at providing safe installation.

Once a project is designed, it must be built. This is accomplished primarily by electrical contractors, their licensed electricians, and helpers. The *National Electrical Code*® includes provisions for the proper installation and construction of electrical work. How to pull wire through conduit, number of wires in a conduit, color coding to be used, and where, how, and under what circumstances the various types of wire and cable may be used are just an example of the many areas of concern.

The supplies and equipment used in an electrical installation must meet minimum requirements that are enumerated in the Code. They must also meet a minimum set of standards, such as color coding and markings on wire and cable, thickness and gauge of materials, and voltage ratings. Hence the manufacturers are constantly referring to the *NEC*® to ensure that their products meet the minimum standards set forth and therefore are saleable.

When work is finished, before being used, it must be inspected by the enforcing authority of the Code. It is at this point that many differences in interpretation emerge. Naturally, it would be nice if we could all come up with one meaning or intent for a given set of rules. Unfortunately, this is not always the case. The inspector must be as familiar with the Code as any other person using it and in some cases more so. He may become the final arbiter before a judge's decision.

Finally, let us not forget those who will take our place, the students and their teachers. The *National Electrical Code*® becomes a useful instructional tool and therefore must be understood and used by the instructor and learned by the student.

It has been found that people using it need to refer to certain topics in the *National Electrical Code*® more frequently than other topics. Some are of concern on a daily basis, whereas others may appear infrequently and then only when a special project is in progress. A copy of the complete *National Electrical Code*® is a necessity and should always be available to the reader. However, a book containing a discussion of only those paragraphs or parts of paragraphs that occur most frequently, and one that is more convenient to carry than the entire Code, will obviously be an excellent aid for daily use. To add to the ease of use of this handy reference, it should be of a size that could easily be carried in a pocket and not take up space in a case. This *Vest Pocket Guide to the National Electrical Code®* was developed and written to fill that need and provide that convenience.

The topics and parts of articles, sections, and paragraphs of the *NEC®* chosen for this book are those that the author felt would be most useful to the reader on a daily basis. Less frequently used parts of the *NEC®* are noted by Article headings for reference purposes only. The material presented is not intended to appear as official interpretations or reproductions of the *NEC®* by the author. Official or formal interpretations may be obtained only from the NFPA and through a definite procedure outlined in the Code. It is also recommended that the authorities having jurisdiction be consulted. This book refers to the 1996 edition of the *National Electrical Code®*.

I would like to dedicate this book to my wife, *Ruth,* and children, *Michael* and *Cheryl, Cathi* and *Marc,* and *Wendy,* and grandchildren, *Brian, Julia, Jessica,* and *Sarah* for their encouragement and patience during its writing.

Marvin J. Fischer

Vest Pocket Guide to the National Electrical Code®

1

GENERAL

Chapter 1 consists of two articles (100 and 110). Before discussing this chapter a few comments are necessary concerning Article 90, Introduction, which precedes Chapter 1. Article 90 covers the purpose of the Code, the scope and arrangement of the Code, enforcement, mandatory rules and explanatory material, the method of obtaining formal interpretations, examination of equipment for safety, wiring planning, and metric units. Section 90-1, Purpose, indicates that the *National Electrical Code*® is *not* intended to be used as a design specification. It contains only provisions for safety. Compliance with the Code does not ensure that the installation is adequate for good service, or that it is convenient or efficient. These criteria are left to other qualified persons, such as electrical designers or engineers.

In certain locations in the code material is extracted from other NFPA documents. This is noted by a superscript letter "x." Appendix A in the code has the source documents identified.

ARTICLE 100
DEFINITIONS

This article contains definitions. Other definitions may be found in sections for specific installations, such as Section 517-2

(Health Care Facilities) and Section 550-2 (Mobile Homes and Mobile Home Parks). There are many others.

ARTICLE 110
REQUIREMENTS FOR ELECTRICAL INSTALLATIONS

A. General

110-2 Approval

110-3 Examination, Identification,Installation, and Use of Equipment

110-4 Voltages

110-5 Conductors

Copper is to be used unless otherwise specified. Sizes given are for copper unless otherwise specified.

110-6 Conductor Sizes

110-7 Insulation Integrity

110-8 Wiring Methods

110-9 Interrupting Rating

110-10 Circuit Impedance and Other Characteristics

The entire electrical circuit, including all devices and wire, must be chosen and coordinated so that a fault will be cleared before extensive damage takes place.

110-11 Deteriorating Agents

110-12 Mechanical Execution of Work

110-13 Mounting and Cooling of Equipment

110-14 Electrical Connections

110-16 Working Space about Electrical Equipment (600 Volts, Nominal or Less)

(a) Clearances for three conditions are noted. Condition 1 consists of live exposed parts on one side and no live or grounded parts on the other side of the working space or guarded live parts (insulating material used) on both sides of the working space. Condition 2 consists of exposed live parts on one side and grounded parts on the other side. Condition 3 consists of exposed live parts on both sides unguarded with the operator between. For voltages to ground between 0 and 150 V, a minimum clear distance of 3 ft. is required in all cases. For voltages to ground between 151 and 600 V, Type 1 requires 3 ft. clearance. Type 2 requires 3½ ft., and Type 3 requires 4 ft. No working space is required behind assemblies where no renewable parts are on the back and all wiring is accessible from the front (such as panelboards). A minimum of 30 in. in required in front of electrical equipment as well as room for a 90° opening of doors. There are three exceptions.

(e) Headroom of 6½ ft. is required above service equipment, motor control centers, panelboards, or switchboards, with the exception of dwelling units not in excess of 200 A. If the equipment exceeds 6½ ft., headroom shall be at least the height of the equipment.

110-17 Guarding Live Parts (600 Volts, Nominal or Less)

110-18 Arcing Parts

110-19 Light and Power from Railway Conductors

110-21 Marking

110-22 Identification of Disconnecting Means

B. Over 600 Volts, Nominal

110-30 General

110-31 Enclosure for Electrical Installations

110-32 Work Space about Equipment

110-33 Entrance and Access to Work Space

110-34 Work Space and Guarding

110-40 Temperature Limitations at Terminations

2

WIRING DESIGN AND PROTECTION

ARTICLE 200
USE AND IDENTIFICATION OF GROUNDED CONDUCTORS

200-1 Scope

Article 200 covers the identification of terminals, grounded conductors in premises wiring systems, and identification of grounded conductors.

200-2 General

The wiring systems of all premises must have a grounded conductor that is identified as stated in Section 200-6. Exceptions to this are noted in Sections 210-10, 215-7, 250-3, 250-5, 250-7, 503-13, 517-63, 668-11, 668-21, and 690-41 Exception. When the grounded conductor is insulated, the insulation must be usable on ungrounded conductors in the same circuits where the voltage is less than 1 kV or on impedance grounded neutral systems of 1 kV and over. For solidly grounded neutral systems of 1 kV or more [Section 250-152(a)], the insulation rating must be 600 V or more.

200-3 Connection to Grounded Systems

Premises Wiring. When the premises wiring system has a grounded conductor, it must be connected to a supply system which also has a grounded conductor.

200-6 Means of Identifying Grounded Conductors

(a) A white or gray covering or insulation is required on insulated grounded conductors of size No. 6 or smaller. MI cable must have distinctive marking at the terminations. Other exceptions are noted.

(b) When the insulated grounded conductor is larger than No. 6, it can be identified the same as conductors smaller than No. 6 or have white markings at its terminations. Requirements for multiconductor flat cable No. 4 or larger are noted.

In both cases (a) and (b), distinctive white markings at the conductor terminations may be used instead of the white or gray coverings if qualified persons perform the maintenance. Exceptions are noted.

(c) When the grounded conductor is in a flexible cord, it must have a gray or white cover or be identified in accordance with Section 400-22.

(d) Requirements for grounded conductors of different systems.

200-7 Use of White or Natural Gray Color

White or gray continuous coverings on a conductor or white or gray markings on terminations can be used only to identify the grounded conductor. These colors may be used for underground conductors: (1) where they are recolored at each termination to show the intended use, (2) for single pole three- and four-way switch loops to supply the switch, (3) for ungrounded conductors in flexible cords, and (4) in circuits of less than 50 V when permitted by Section 250-5(a).

200-9 Means of Identification of Terminals

200-10 Identification of Terminals

When conductors are to be attached to more than “one side” of a circuit, they must have the device terminals identified. This is not necessary when the connection is evident, in lighting and appliance branch-circuit panelboards, and in devices rated over 30 A—other than polarized attachment plugs and receptacles.

Polarized attachment plugs and receptacles must have the terminal for the grounded conductor marked white. Refer to Section 215-219 for equipment grounding terminals. The terminal for the grounded conductor in screw-shell devices is the one connected to the screw shell. If there is a lead connected to the screw shell, it must be white or gray.

200-11 Polarity of Connections

ARTICLE 210
BRANCH CIRCUITS

A. General Provisions

210-1 Scope

Article 210 applies to all branch circuits except motor loads. When a branch circuit supplies this type of load in combination with a motor load, both this article and Article 430 must be followed. When the branch circuit supplies only motor loads, Article 430 must be followed.

210-2 Other Articles for Specific-Purpose Branch Circuits

The rules for branch circuits in the table on page 7 amend or supplement this article as noted.

210-3 Rating

Branch circuits are rated according to the maximum permitted rating or setting of their overcurrent device. Generally, they are classified as 15, 20, 30, 40, and 50 A.

210-4 Multiwire Branch Circuits

Multiwire branch circuits are permitted. They can only supply line to neutral load, except if it supplies one piece of equipment or if all ungrounded conductors of the circuit are opened by the overcurrent device simultaneously. There is a dwelling unit restriction.

210-5 Color Code for Branch Circuits

In general, the grounded conductor must have a continuous white or gray cover. When different systems are installed in the same enclosure, one grounded conductor must have a white covering. The other system grounded conductors must have a white cover with a

	Article	Section
Air-Conditioning and Refrigerating Equipment		440-6
		440-31
		440-32
Busways		364-9
Circuits and Equipment Operating at Less than 50 Volts	720	
Central Heating Equipment Other than Fixed Electric Space Heating Equipment		422-7
Class 1, Class 2, and Class 3 Remote-Control, Signaling, and Power-Limited Circuits	725	
Closed-Loop and Programmed Power Distribution Systems	780	
Cranes and Hoists		610-42
Electronic Computer/Data Processing Equipment		645-5
Electronic Signs and Outline Lighting		600-6
Electric Welders	630	
Elevators, Dumbwaiters, Escalators, Moving Walks, Wheel Chair Lifts, and Stairway Chair Lifts		620-61
Fire Protective Signaling Systems	760	
Fixed Electric Space Heating Equipment		424-3
Fixed Outdoor Electric De-icing and Snow-Melting Equipment		426-4
Induction and Dielectric Heating Equipment	665	
Infared Lamp Industrial Heating Equipment		422-15
		424-3
Marinas and Boatyards		555-4

	Article	*Section*
Mobile Homes and Mobile Home Parks	550	
Motion Picture and Television Studios and Similar Locations	530	
Motors, Motor Circuits, and Controllers	430	
Pipe Organs		650-6
Recreational Vehicles and Recreational Vehicle Parks	551	
Sound-Recording and Similar Equipment		640-6
Switchboards and Panelboards	384-22	
Theaters, Audience Areas of Motion Picture and Television Studios, and Similar Locations		520-41
		520-52
		520-62
X-Ray Equipment		660-2
		517-143

stripe of another color, except green. MI cables are marked at the terminals. When only qualified persons perform maintenance and supervision of the installations, distinctive white markings may be used at the terminals in lieu of the white or gray coverings.

210-6 Branch-Circuit Voltage Limitations

(a) The voltage between conductors cannot be greater than 120 V for lighting fixtures and cord-and-plug connected loads of less than 1440 VA or less than ¼ hp in dwelling units, hotel and motel guest rooms, and the like.

(b) Circuits not exceeding 120 V to ground can supply auxiliary equipment of electric discharge lamps, utilization equipment, and terminals of lampholders within their voltage rating.

(c) Circuits between 120 and 277 V to ground can supply mogul-base screw-shell lampholders, lampholders used within their voltage rating, auxiliary equipment to electric discharge lamps, listed electric discharge fixtures, and utilization equipment. Additionally, these circuits can supply listed lampholders that are supplied at 120 V or less from a stepdown autotransformer which is part of the fixture and the outer shell terminal is connected to the grounded conductor of the branch circuit.

(d) 600 V between conductors.

210-7 Receptacles and Cord Connectors

In general, 15- and 20-A branch circuits are to have grounding type receptacles installed. Exceptions, methods, and replacements are listed.

210-8 Ground-Fault Protection for Personnel

(a) Dwelling Units. All 125-V single-phase 15- and 20-A receptacles must have ground fault protection for personnel when installed in bathrooms, garages and grade level parts of storage or work areas of unfinished accessory buildings (two exceptions), outdoors (one exception), crawl spaces at or below grade level, unfinished basements (two exceptions), kitchens where the receptacles are serving counter tops, and wet bar sinks where the receptacle is within 6 ft. of the edge of the sink and to serve the counter top.

(b) Other than Dwelling Units. Ground-fault circuit interrupter receptacles are required in bathrooms and on roofs on all 125-V single-phase 15- and 20-A outlets.

210-9 Circuits Derived from Autotransformers

210-10 Ungrounded Conductors Tapped from Grounded Systems

B. Branch-Circuit Ratings

210-19 Conductors—Minimum Ampacity and Size

(a) General. Generally, conductors must have a capacity not less than the maximum load. Conductors for multioutlet branch circuits for receptacles must be rated at least as that of the branch circuit. Various tables are supplied. If the conductors are sized so that the maximum voltage drop on the branch circuit to the outlet is not greater than 3% of the total voltage drop for both feeder and branch circuit is not greater than 5%, a reasonable efficiency of operation will be provided.

(b) Household Ranges and Cooking Appliances. Ratings cannot be less than the rating of the load. In cases of ranges of 8¾ kW or more, the minimum circuit rating is 40 A. Neutrals and conductors are permitted to be smaller in accordance with the Code.

(c) Other Loads. Ratings must be sufficient for the loads supplied, but in no case can a conductor be smaller than No. 14. Exceptions are made for tap conductors and fixture wires and cords.

210-20 Overcurrent Protection

The equipment and conductors used on branch circuits must be protected by overcurrent devices. The ratings or settings are set forth in Section 240-3 for conductors, Section 200-2 for equipment, and Section 210-21 for outlet devices. An exception is made for tap conductors, fixture wire, and cords which are permitted in Section 210-19(c) and 240-4. These are protected by the branch-circuit protection.

210-21 Outlet Devices

(a) Lampholders connected to a branch circuit rated at more than 20 A must be a heavy-duty type.

(b) Receptacles. A single receptacle can have a rating only equal to or larger than the rating of the individual branch circuit it is on except when installed in accordance with Section 430-81(c). The maximum load from a cord and plug connection where there is more than one receptacle on the circuit is shown in Table 210-21(b)(2). The ratings of receptacles where there is more than one on a circuit are shown in Table 210-21(b)(3). If the branch circuit rating is more than 50 A, then the receptacle rating must be at least as large as the branch circuit. The ampere rating of a range receptacle can be based on a single range as noted in Table 220-19. There is an exception for cord- and plug-connected arc welders.

TABLE 210-21(b)(2) Maximum Cord- and Plug-Connected Load to Receptacle

Circuit Rating Amperes	*Receptacle Rating Amperes*	*Maximum Load Amperes*
15 or 20	15	12
20	20	16
30	30	24

TABLE 210-21(b)(3) Receptacle Ratings for Various Size Circuits

Circuit Rating Amperes	*Receptacle Rating Amperes*
15	Not over 15
20	15 or 20
30	30
40	40 or 50
50	50

210-22 Maximum Loads

The total load cannot exceed the rating of the branch circuit and the loads noted in Sections 210-22(a) through (c). Article 430 governs only motor-operated loads on a circuit. Article 440 governs only air conditioning and refrigeration equipment on a circuit. If the circuit combines motor-operated loads that are fastened in place and larger than ⅛ hp, the total load is 125% of the largest motor plus the sum of all other loads. When the circuit contains inductive lighting loads supplied by ballasts, autotransformers, or transformers, the actual ampere ratings should be used. Continuous loads must have the branch circuit computed load equal to or more than the noncontinuous load plus 125% of the continuous load. This shall also be the minimum branch circuit ampacity of the conductor size. An exception is made for overcurrent devices and assemblies that are listed for 100% rating continuous operation.

210-23 Permissible Loads

The load cannot exceed the branch circuit rating. Branch circuits that supply more than one receptacle or outlet follow the rules in (a) through (d) below and summarized in Section 210-24 and Table 210-24.

(a) 15- and 20-A branch circuits can supply lighting loads, utilization equipment, or a combination. The rating of one cord- and plug-connected utilization equipment cannot be more than 80% of the branch-circuit ratings. Connected utilization equipment cannot be more than 50% of the branch-circuit rating where both connected and cord-and-plug loads are used.

(b) 30-A branch circuits can supply heavy-duty lampholders in fixed lighting units in all occupancies except dwelling units. They can also supply utilization equipment in all occupancies. One cord- and plug-connected utilization equipment cannot be more than 80% of the branch circuit rating.

(c) 40- and 50-A branch circuits supply the same lighting loads as the 30-A branch circuit, connected cooking appliances in any occupancy, and infrared heating units in occupancies dwelling units. They can supply fastened-in-place cooking appliances in dwelling units.

(d) If the branch circuit is larger than 50 A, it can only supply nonlighting loads.

210-24 Branch-Circuit Requirements—Summary

Table 210-24 provides a summary of branch-circuit requirements other than those required in Section 220-4(b).

210-25 Common Area Branch Circuits

Dwelling unit branch circuits can only supply circuits in or associated with that dwelling unit in multifamily dwellings. Branch circuits supplying loads such as lighting, alarms, and signals for common areas cannot be supplied from a dwelling unit panelboard.

C. Required Outlets

210-50 General

A cord connection is considered a receptacle outlet if it is supported by a permanently installed cord pendant. When it is known that a flexible cord will be used, a receptacle outlet should be installed. Appliance outlets are to be installed within 6 ft of the location of the appliance.

210-52 Dwelling Unit Receptacle Outlets

(a) General Provisions. Receptacle outlets must be installed in dwelling units so that a receptacle outlet is not more than 6 ft. from any point in a wall space along the floor line. This applies to wall spaces 2 ft. wide or larger, those containing fixed panels in exterior walls, and those containing fixed room dividers (i.e. railings or free standing bar-type counters). Sliding panels in exterior walls are excluded. A wall space is an unbroken wall along the floor (unbroken by doors, fireplaces, etc.). Each wall space 2 ft. wide or more is considered separately. However, more than one wall, if connected, is considered one wall space, such as around a corner. Receptacles are to be spaced equally. Receptacles in floors and in lighting fixtures or appliances 5½ ft. above the floor are not counted. The areas that this applies to are kitchen, family room, dining room, living room, parlor, library, den, sunroom, bedroom, recreation room, and similar types of rooms.

(b) Small Appliances. The required [Section 220-4(b)] two or more 20-A small appliance branch circuits must serve all receptacles covered by Sections 210-52(a) and (c) in areas such as dining room, breakfast room, pantry, and kitchen, including refrigeration equipment. Exceptions are made for switched receptacles supplied from a general purpose branch circuit as described in Section 210-70(a), Exception 1 and refrigeration equipment can be supplied from a 15-A or greater individual branch circuit. No other outlets are allowed on the required two or more 20-A small appliance branch circuits except clocks and power for supplemental equipment on gas fired ranges, ovens, or counter mounted cooking equipment.

(c) Countertops. A receptacle is to be installed in each wall countertop space wider than 12 in. in kitchens and dining areas. No point along the wall can be farther than 24 in. from a receptacle measured horizontally. If an island or peninsular countertop is 24 in. or more in one dimension and 12 in. or more in the other dimension it must have at least one receptacle. If a counter top is separated by a range, sink, or refrigerator each part is to be considered a separate counter top. No receptacle is to be more than 18 in. above the counter top. Receptacles are not to be installed face up. A receptacle which is covered by

TABLE 210-24 Summary of Branch-Circuit Requirements

Circuit Rating	*15 Amp*	*20 Amp*	*30 Amp*	*40 Amp*	*50 Amp*
Conductors: (min. size)					
Circuit wires*	14	12	10	8	6
Taps	14	14	14	12	12
Fixture wires and cords			Refer to Section 240-4		
Overcurrent Protection	*15 Amp*	*20 Amp*	*30 Amp*	*40 Amp*	*50 Amp*
Outlet devices:					
Lampholders permitted	Any type	Any type	Heavy duty	Heavy duty	Heavy duty
Receptacle rating†	15 Max. Amp	15 or 20 Amp	30 Amp	40 or 50 Amp	50 Amp

Maximum Load:	*15 Amp*	*20 Amp*	*30 Amp*	*40 Amp*	*50 Amp*
Permissible load	Refer to Section 210-23(a)	Refer to Section 210-23(a)	Refer to Section 210-23(b)	Refer to Section 210-23(c)	Refer to Section 210-23(c)

* These gauges are for copper conductors.
† For receptacle rating of cord-connected electric-discharge lighting fixtures, see Section 410-30(c).

a fixed appliance or one in a dedicated space is not to be counted as a required outlet. There is an exception for the handicapped and a special island and peninsula condition.

(d) Bathrooms. A minimum of one wall receptacle is to be installed next to each basin. Bathroom receptacles should have at least one 20-A circuit. This can supply only those outlets. Refer to section 210-8(a)(1). Receptacles cannot be mounted face up on counter tops.

(e) Outdoor Outlets. A minimum of one outdoor outlet is to be installed outdoors for a one-family house. A two-family house must have at least one outdoor outlet for each unit that is at grade level. These receptacles are to be installed in the front and back. The receptacle outlet cannot be higher than 6½ ft. above grade.

(f) Laundry Areas. A minimum of one outlet is to be installed in the laundry area. This is not required in multifamily dwelling units where laundry rooms are provided or laundry facilities are not individually installed.

(g) Basements and Garages. A minimum of one receptacle is to be installed in the basement and garage of a one-family dwelling. Refer to Sections 210-8(a)(2) and (a)(4).

(h) Hallways. One receptacle must be installed in a hallway 10 ft. or longer. The length of the hallway is measured along the centerline of the hall without going through a doorway.

210-60 Guest Rooms

Guest rooms in a hotel or motel must comply with Section 210-52 unless otherwise required for permanently installed furniture. Refer to Section 210-8(b). There is an exception for permanent furniture layout.

210-62 Show Windows

A minimum of one receptacle is required for each 12 linear feet or major fraction of a show window. It is to be measured at the maximum width and the receptacle to be installed above the window.

210-63 Heating, Air-Conditioning, and Refrigeration Equipment Outlet

210-70 Lighting Outlets Required

(a) Dwelling Units. Every habitable room, as well as outdoor entrances and exits, attached garages, detached garages with electric power, hallways, stairways, and bathrooms, must have a minimum of one lighting outlet controlled by a wall switch. A receptacle is acceptable in place of the lighting outlet in habitable rooms other than kitchens and bathrooms. Remote automatic or central control can be used in hallways, outdoor entrances, and stairways. When an attic, utility room, underfloor space, or basement is used for storage or equipment requiring service, it must have a minimum of one lighting outlet controlled at the point of entry. There are requirements for stair lighting. The lighting outlets can be controlled by an occupancy sensor if it is in addition to a wall switch or is at the location of the wall switch with a manual override.

(b) Guest Rooms. Guest rooms in hotels, motels, and similar installations must have a minimum of one lighting outlet or receptacle connected to a wall switch.

(c) Other Locations. Requirements for lighting in attics or underfloor spaces for heating, air-conditioning, refrigeration equipment, or other equipment needing servicing.

ARTICLE 215
FEEDERS

215-1 Scope

215-2 Minimum Rating and Size

If a three-wire feeder supplies more than two two-wire branch circuits or two or more three-wire branch circuits, it must be sized for at least 30 A. The same is true of a two-wire feeder supplying two or more two-wire branch circuits and two or more four-wire branch circuits supplied by a three-phase, four-wire feeder. If a feeder carries the total load of the service entrance conductors and the ampacity of the service entrance conductor is 55 A or less, the feeder must be sized at least the same.

215-3 Overcurrent Protection

Refer to Article 240, Part a.

215-4 Feeders with a Common Neutral

215-5 Diagram of Feeders

215-6 Feeder Conductor Grounding Means

If equipment grounding conductors are required, refer to Section 250-57.

215-7 Ungrounded Conductors Tapped from Grounded Systems

Permitted with a switching device with a pole in each ungrounded conductor.

215-8 Means of Identifying Conductor with Higher Voltage to Ground

215-9 Ground-Fault Circuit-Interrupter Protection for Personnel

When a feeder supplies 15- and 20-A receptacle branch circuits, ground-fault circuit interrupter may be installed in the feeder instead of the requirements of Section 210-8 and Article 305.

215-10 Ground-Fault Protection of Equipment

Ground-fault protection of equipment in accordance with Section 230-95 must be provided for a feeder in a solidly grounded wye system of 150 to 600 V to ground and a disconnect rated 1000 A or more unless it is provided on the supply side of the feeder.

215-11 Circuits Derived from Autotransformers

ARTICLE 220 BRANCH-CIRCUIT, FEEDER, AND SERVICE CALCULATIONS

A. General

220-1 Scope

220-2 Voltages

220-3 Computation of Branch Circuits

(a) Continuous and Noncontinuous Loads. The rating of the branch circuit is to be at least that of the noncontinous load plus 125% of the continuous load. The minimum conductor size shall have an ampacity equal to this without any adjustments.

(b) Lighting Load for Listed Occupancies. Table 220-3(b) lists the minimum load required based on floor area, which is calculated using outside dimensions. Where an area cannot be used in dwelling units (such as a garage or open porch), it does not have to be included. It should be noted that these values are a minimum and may not provide for an adequate installation.

(c) Other Loads: All Occupancies. In occupancies not listed in Table 220-3(b), all receptacle outlets on one strap must be rated at a minimum of 180 V/A. Outlets for heavy-duty lamp-

TABLE 220-3(b) General Lighting Loads by Occupancies

Type of Occupancy	*Unit Load per Sq. Ft. (Volt-Amperes)*
Armories and auditoriums	1
Banks	3½*
Barber shops and beauty parlors	3
Churches	1
Clubs	2
Court rooms	2
Dwelling units†	3
Garages—commercial (storage)	½
Hospitals	2
Hotels and motels, including apartment houses without provisions for cooking by tenants†	2
Industrial commercial (loft) buildings	2
Lodge rooms	1½
Office buildings	3½*
Restaurants	2
Schools	3
Stores	3
Warehouses (storage)	¼
In any of the above occupancies except one-family dwellings and individual dwelling units of two family and multi-family dwelling units:	
Assembly halls and auditoriums	1
Halls, corridors, closets, stairways	½
Storage spaces	¼

For SI units: one square foot = 0.093 square meter.

* In addition a unit load of 1 VA per square foot shall be included for general-purpose receptacle outlets when the actual number of general-purpose receptacle outlets is unknown.

† All general use receptacle outlets of 20-A or less rating in one-family, two-family, and multifamily dwellings and in guest rooms of hotels and motels [except those connected to the receptacle circuits specified in Section 220-4(b) and (c)] shall be considered as outlets for general illumination, and no additional load calculations shall be required for such outlets.

holders are to be rated at 600 V/A. Outlets for specific loads must be the rating of the load. Motor loads are covered in Sections 430-22 and 430-24 and in Article 440. The load for track lighting is noted in Section 410-102.

Multioutlet assemblies need only be counted as one outlet (11/2 A) for each 5-ft. length unless many appliances are used simultaneously or, where many appliances are likely to be connected, each 1-ft. length. Table 220-19 can be used for loads for household electric ranges. Show-window lighting can be calculated as 200 VA per foot. Outlets for frames and switchboards in telephone exchanges are not covered by this section. Demand factors in Section 220-18 can be used for household electric dryers. Receptacles specified in Section 220-4(b) and (c) are not bound by these provisions. Sign and outline lighting are to be 1200 VA for each branch circuit noted in Section 600-5(a).

(d) Loads for Additions to Existing Installations. Loads for additions to dwelling units greater than 500 square feet are to be calculated in accordance with Section 220-3(b). Additions to circuits should be calculated as noted in Section 220-3(b) and (c). Loads for other than dwelling units are to be calculated as noted in Section 220-3(b) and (c).

220-4 Branch Circuits Required

(a) Number of Branch Circuits. The number of branch-circuits is to be calculated from the total load and should be capable of supplying the lighting and appliances. The load on any circuit cannot be more than that designated in Section 210-22.

(b) Small-Appliance Branch Circuits—Dwelling Unit. In addition to (a), at least two 20-A small appliance branch circuits are to be installed for the small appliance loads required by Section 210-52.

(c) Laundry Branch Circuits—Dwelling Unit. At least one 20-A branch circuit with no other outlets on it must be installed for laundry receptacles.

(d) Load Evenly Proportioned among Branch Circuits. The load calculated using Table 220-3(b) is to be divided as evenly as possible among all branch circuits.

B. Feeders and Services

220-10 General

(a) Ampacity and Computed Loads. Feeders must be able to supply the required load. They must be sized to supply the connected load with applied allowable demand factors.

(b) Continuous and Noncontinuous Loads. For a continuous load or combination of continuous and noncontinuous loads, the rating of the overcurrent device and ampacity of the feeder conductor must be at least 125% of the continuous load plus the noncontinuous load. The 125% factor can be reduced to 100% where the article is rated for continuous operation.

220-11 General Lighting

To calculate the general lighting load, use the demand factors in Table 220-11.

220-12 Show-Window Lighting

Use at least 200 VA per linear foot.

220-13 Receptacle Loads—Nondwelling Units

Receptacle loads may either be lumped in with lighting loads and demand factors from Table 220-11 or figured separately and demand factors from Table 220-13 used. The load is to be figured at not more than 180 V/A per receptacle in accordance with Section 220-3(c)(6).

220-14 Motors

Calculate as directed by Sections 430-24, 430-25, and 430-26.

220-15 Fixed Electric Space Heating

Generally, the load is to be calculated at 100% of the connected load. If, however, the total load is less as a result of intermittent operation or diversity, the authorities can reduce the calculated load. Optional calculations as shown in Sections 220-30 and 220-31 can be used in a dwelling unit and as shown in Section 220-32 for multifamily units. The feeder load current rating can never be less than that of the largest branch circuit it supplies.

220-16 Small Appliance and Laundry Loads—Dwelling Units

Use 1500 VA for each small-appliance and laundry branch circuit called for in Section 220-4(b) and (c). This can be included in the general lighting load, and demand factors of Table 220-11 applied.

220-17 Appliance Load—Dwelling Units

A demand factor of 75% can be applied to nameplate ratings of four or more appliances in a dwelling. This does not apply to

TABLE 220-11 Lighting Load Feeder Demand Factors

Type of Occupancy	*Portion of Lighting Load to Which Demand Factor Applies (Volt-Amperes)*	*Demand Factor Percent*
Dwelling units	First 3000 or less at	100
	From 3001 to 120,000 at	35
	Remainder over 120,000 at	25
Hospitals*	First 50,000 or less at	40
	Remainder over 50,000 at	20
Hotels and motels—including apartment houses without provision for cooking by tenants*	First 20,000 or less at	50
	From 20,001 to 100,000 at	40
	Remainder over 100,000 at	30
Warehouses (storage)	First 12,500 or less at	100
	Remainder over 12,500 at	50
All others	Total volt-amperes	100

* The demand factors of this table shall not apply to the computed load of feeders to areas in hospitals, hotels, and motels where the entire lighting is likely to be used at one time, as in operating rooms, ballrooms, or dining rooms.

TABLE 220-13 Demand Factors for Nondwelling Receptacle Loads

Portion of Receptacle Load to Which Demand Factor Applies (volt-amperes)	*Demand Factor Percent*
First 10 kVA or less	100
Remander over 10 kVA at	50

electric ranges, clothes dryers, and space heating or air-conditioning equipment.

220-18 Electric Clothes Dryers—Dwelling Units

Use a load of 5000 VA or the nameplate rating, whichever is larger. Demand factors listed in Table 220-18 can be used.

TABLE 220-18 Demand Factors for Household Electric Clother Dryers

Number of Dryers	*Demand Factor Percent*
1	100
2	100
3	100
4	100
5	80
6	70
7	65
8	60
9	55
10	50
11-13	45
14-19	40
20-24	35
25-29	32.5
30-34	30
35-39	27.5
40 and over	25

220-19 Electric Ranges and Other Cooking Appliances—Dwelling Unit(s)

This section applies to household electric ranges, wall-mounted ovens, counter-mounted cooking units, and other household cooking units larger than 1¾ kW. For those units the demand load in Table 220-19 can be used. kVA and kW are considered equivalent for this section.

For single-phase ranges on a three-phase four-wire feeder, use twice the number of units between two phases. Reference to the code should be made for five notes to Table 220-19 concerning ranges between 12 and 27 kW and between 1¾ and 8¾ kW for alternate allowable means of calculation.

220-20 Kitchen Equipment—Other Than Dwelling Units

This section applies to commercial cooking equipment, dishwasher booster heaters, water heaters, and other kitchen equipment. Demand factors shown in Table 220-20 can be used. They apply to thermostatic controlled or intermittently used kitchen equipment, not air conditioning, space heating or ventilating equipment. The feeder must always be larger than the sum of the two largest kitchen equipment loads.

220-21 Noncoincident Loads

If two loads on the same feeder will not be operating simultaneously, the smaller of the two can be omitted for calculation purposes.

220-22 Feeder Neutral Load

The feeder neutral load is by definition the maximum unbalanced load between the neutral and one ungrounded conductor. In three-wire two-phase or five-wire two-phase systems the load is then multiplied by 140%. A factor of 70% is used in the case of household electric ranges, wall-mounted ovens, electric dryers, and counter-mounted cooking units. A factor of 70% may also be used for load more than 200 A in systems with three-wire dc or single-phase ac; three-phase, four-wire; two-phase, three-wire; and two-phase, five-wire. Reduction of neutral capacity in a four-wire wye-connected three-phase system is not permitted for nonlinear loads.

TABLR 220-19 Demand Loads for Household Electric Ranges, Wall-Mounted Ovens, Counter-Mounted Cooking Units, and Other Household Cooking Appliances over $1\frac{3}{4}$ kW Rating

	Maximum Demand (See Notes)	*Demand Factors Percent (See Note 3)*	
Number of Appliances	*Column A (Not over 12 kW Rating)*	*Column B (Less Than $3\frac{1}{2}$ kW Rating)*	*Column C ($3\frac{1}{2}$ kW to $8\frac{3}{4}$ kW Rating)*
1	8 kW	80	80
2	11 kW	75	65
3	14 kW	70	55
4	17 kW	66	50
5	20 kW	62	45
6	21 kW	59	43
7	22 kW	56	40
8	23 kW	53	36
9	24 kW	51	35
10	25 kW	49	34
11	26 kW	47	32
12	27 kW	45	32
13	28 kW	43	32

TABLE 220-19 (Continued)

14	29 kW	41	32
15	30 kW	40	32
16	31 kW	39	28
17	32 kW	38	28
18	33 kW	37	28
19	34 kW	36	28
20	35 kW	35	28
21	36 kW	34	26
22	37 kW	33	26
23	38 kW	32	26
24	39 kW	31	26
25	40 kW	30	26
26-30	15 kW plus 1 kW for each range	30	24
31-40		30	22
41-50	25 kW plus ¾ kW for each range	30	20
51-60		30	18
61 and over		30	16

Column A to be used in all cases except as otherwise permitted in Note 3.

Note 1. Over 12 kW through 27 kW ranges all of same rating. For ranges individually rated more than 12 kW but not more than 27 kW, the maximum demand in Column A shall be increased 5 percent for each additional kW of rating or major fraction thereof by which the rating of individual ranges exceeds 12 kW.

Note 2. Over 8 3/4 kW through 27 kW ranges of unequal ratings. For ranges individually rated more than 8 3/4 kW and of different ratings, but none exceeding 27 kW, an average value of rating shall be computed by adding together the ratings of all ranges to obtain the total connected load (using 12 kW for any range rated less than 12 kW) and dividing by the total number of ranges; and then the maximum demand in Column A shall be increased 5 percent for each kW or major fraction thereof by which this average value exceeds 12 kW.

Note 3. Over 1 3/4 kW through 8 3/4 kW. In lieu of the method provided in Column A, it shall be permissible to add the nameplate ratings of all household cooking appliances rated more than 1 3/4 kW but not more than 8 3/4 kW and multiply the sum by the demand factors specified in Column B or C for the given number of appliances.

Note 4. Branch-Circuit Load. It shall be permissible to compute the branch-circuit load for one range in accordance with Table 220-19. The branch-circuit load for one wall-mounted oven or one counter-mounted cooking unit shall be the nameplate rating of the appliance. The branch-circuit load for a counter-mounted cooking unit and not more than two wall-mounted ovens, all supplied from a single branch circuit and located in the same room, shall be computed by adding the nameplate rating of the individual appliances and treating this total as equivalent to one range.

Note 5. This table also applies to household cooking appliances rated over 1 3/4 kW and used in instructional programs.

(FPN No. 1): See Table 220-20 for commercial cooking equipment.

(FPN No. 2): See Examples, Chapter 9.

TABLE 220-20 Feeder Demand Factors for Kitchen Equipment—Other Than Dwelling Unit(s)

Number of Units of Equipment	*Demand Factors Percent*
1	100
2	100
3	90
4	80
5	70
6 and over	65

C. Optional Calculations for Computing Feeder and Service Loads

220-30 Optional Calculation—Dwelling Unit

Table 220-30 can be used as an optional method for calculating loads in a dwelling unit with a minimum load of 100 A and a

TABLE 220-30 Optional Calculation for Dwelling Unit Load in kVA

Largest of the Following Five Selections:

(1) 100 percent of the nameplate rating(s) of the air conditioning and cooling, including heat pump compressors.

(2) 100 percent of the nameplate ratings of electric thermal storage and other heating systems where the usual load is expected to be continuous at the full nameplate value. Systems qualify under this selection shall not be figured under any other selection in this table.

(3) 65 percent of the nameplate rating(s) of the central electric space heating including integral supplemental heating in heat pumps.

(4) 65 percent of the nameplate rating(s) of electric space heating if less than four separately controlled units.

(5) 40 percent of the nameplate rating(s) of electric space heating of four or more separately controlled units.

Plus: 100 percent of the first 10 kVA of all other load.
Plus: 40 percent of the remainder of all other load.

single service of three-wire 120/240 V and 120/208Y V. Other loads and remainder of other loads in the table include 1500 VA for small appliance and laundry circuits, 3 VA per square foot for general lighting, and nameplate ratingsof permanently connected, located on a specific circuit, or fastened-in-place appliances, wall-mounted ovens, motors, low-power-factor loads, and countertop cooking units.

220-31 Optional Calculations for Additional Loads in Existing Dwellings

This section applies only to existing 120/240-V and 120/208Y-V three-wire services. It gives an alternate means of calculating loads. Refer to the section in the Code.

220-32 Optional Calculation—Multifamily Dwelling

This section permits an alternate means of calculation of feeder and service loads when each dwelling is supplied by a maximum of one feeder, has electric cooking, and has electric space heating or air conditioning. The alternate calculation is based on the demand factors shown in Table 220-32. These demand factors apply to the small-appliance branch-circuit load, 3 VA per square foot general load, the nameplate rating of all appliances and motors, and air-conditioning or space heating loads. House loads are calculated in accordance with Part B of this article and are in addition to the dwelling unit loads calculated by Table 220-32.

220-33 Optional Calculation—Two Dwelling Units

220-34 Optional Method—Schools

Refer to Table 220-34.

220-35 Optional Calculations for Additional Loads to Existing Installations

Actual demand data can be used if the maximum readings are known for at least 1 year, the existing demand multiplied by 125% and the new loads are not larger than the wire capacity, and proper feeder and service overcurrent protection are provided, in accordance with sections 240-3 and 230-90, respectively.

TABLE 220-32 Optional Calculation—Demand Factors for Three or More Multifamily Dwelling Units

Number of Dwelling Units	*Demand Factor Percent*
3–5	45
6–7	44
8–10	43
11	42
12–13	41
14–15	40
16–17	39
18–20	38
21	37
22–23	36
24–25	35
26–27	34
28–30	33
31	32
32–33	31
34–36	30
37–38	29
39–42	28
43–45	27
46–50	26
51–55	25
56–61	24
62 and over	23

220-36 Optional Calculations—New Restaurants

Table 220-36 can be used to calculate a service load or a feeder for a new restaurant if the feeder serves the total load. The service-entrance overload protection must be in accordance with Sections 230-90 and 240-3. The feeder and subfeeder conductors do not have to be larger than the service-entrance conductors. If this optional calculation is used, the neutral load can be

TABLE 220-34 Optional Method—Demand Factors for Feeders and Service-Entrance Conductors for Schools

Connected Load Volt-Amperes per Square Foot	*Demand Factors Percent*
The first 3 VA/sq. ft. at	100
Plus, over 3 to 20 VA/sq. ft. at	75
Plus, remainder over 20 VA/sq. ft. at	25

determined by Section 220-22. Table 220-36 indicates demand factors for all-electric and not-all-electric loads. For a connected load of 0 to 325 kVA, the all-electric demand factor is 80% and the non-all-electric demand factor is 100%. For a connected load of over 325 kVA, the all-electric demand factor is 50% and the non-all-electric demand factor is 70%.

D. Method for Computing Farm Loads

220-40 Farm Loads—Buildings and Other Loads

220-41 Farm Loads—Total

Note: Examples for the calculations referred to in Article 220 are shown in Part B of Chapter 9 of the *NEC*®. They are shown in the same location in this book for correlation.

ARTICLE 225
OUTSIDE BRANCH CIRCUITS AND FEEDERS

225-1 Scope

This article covers feeders and branch circuits located outside on premises between buildings or poles, or for equipment on the outside of buildings or poles.

225-2 Other Articles

The following list refers to other articles in the Code which are pertinent to outside branch circuits and feeders.

	Article
Branch Circuits	210
Class 1, Class 2, and Class 3 Remote-Control, Signaling, and Power-Limited Circuits	725
Communications Circuits	800
Community Antenna Television and Radio Distribution Systems	820
Conductors for General Wiring	310
Electrically Driven or Controlled Irrigation Machines	675
Electric Signs and Outline Lighting	600
Feeders	215
Fire Protective Signaling Systems	760
Fixed Outdoor Electric De-icing and Snow-Melting Equipment	426
Floating Buildings	553
Grounding	250
Hazardous (Classified) Locations	500
Hazardous (Classified) Locations, Specific	510
Marinas and Boatyards	555
Messenger Supported Wiring	321
Open Wiring on Insulators	320
Over 600 Volts, General	710
Overcurrent Protection	240
Radio and Television Equipment	810
Services	230
Solar Photovoltaic Systems	690
Swimming Pools, Fountains, and Similar Installations	680
Use and Identification of Grounded Conductors	200

225-3 Calculation of Load

The same as in Article 220.

225-4 Conductor Covering

Conductors must be insulated within 10 ft. of a building. Thermoplastic or rubber insulation must be used in raceways. An exception is made for equipment grounding conductors and grounded circuit conductors.

225-5 Size of Conductors

Size in accordance with Section 310-15. Determine loads in accordance with Section 220-30 and Part B of Article 220.

225-6 Minimum Size of Conductor

(a) Overhead Spans

600 V or less, up to 50 ft.: No. 10 copper, No. 8 aluminum except where supported by a messenger wire
600 V or less, over 50 ft.: No. 8 copper, No. 6 aluminum except where supported by a messenger wire
More than 600 V: No. 6 copper, No. 4 aluminum for open conductors; No. 8 copper, No. 6 aluminum for cable

(b) No. 12 or Larger for Festoon Lighting except where messenger wires support them

225-7 Lighting Equipment Installed Outdoors

225-8 Disconnection

The disconnecting means has to be as stated in Section 240-40. Every building must have its own disconnecting means. There is an exception under certain conditions for large capacity multibuilding industrial installations, those permitted in Article 685 and for poles or groups of poles used for lighting where disconnecting means are remote. The disconnecting means must be good for service equipment. The disconnects have to be installed as noted in Sections 230-71 and 230-72. If more than one feeder, branch circuit, or service supplies a building, a permanent plaque or directory has to be installed at each feeder, branch circuit, or service location noting all other feeder, branch circuit, and service locations feeding the building. There is an exception for large capacity multibuildings under one management under certain conditions.

225-9 Overcurrent Protection

Refer to Section 210-20 and Article 240.

225-10 Wiring on Buildings

225-11 Circuit Exits and Entrances

Refer to Sections 230-52, and 230-54.

225-12 Open Conductor Supports

225-13 Festoon Supports

225-14 Open Conductor Spacings

225-15 Supports over Buildings

Refer to Section 230-29.

225-16 Point of Attachment to Buildings

Refer to Section 230-26.

225-17 Means of Attachment to Buildings

Refer to Section 230-27.

225-18 Clearance from Ground

Open conductors and open multiconductor cables not over 600 volts must conform to the following:

10 ft.: above finished grade: 150 V to ground
12 ft.: over residential driveways and commercial areas not subject to truck traffic: 300 V to ground limitation
15 ft.: same as 12 ft. above: exceeding 300 V to ground
18 ft.: above streets, driveways, etc. subject to truck traffic as well as land used by vehicles such as forest, grazing, etc.

225-19 Clearances from Buildings for Conductors of Not Over 600 Volts, Nominal

(a) Above Roofs. Not less than 8 ft. vertical clearance from the roof surface maintained for a distance of not less than 3 ft. in all directions from the edge of the roof except the final span to the attachment to the building. If subject to pedestrian or vehicular traffic, vertical clearances should be in accordance with Section 225-18. Other exceptions are as follows:

3 ft.: voltage between conductors not more than 300 V and roof slope is maximum of 4 in. in 12 in.
18 in.: voltage between conductors not more than 300 V, not more than 6 ft. of conductor, 4 ft. horizontally, do not pass over an overhang and are terminated in approved support or raceway through the roof
3 ft.: clearance does not apply to the final conductor span attached to the roof

(b) From Nonbuilding or Nonbridge Structures. Clearances in all directions to be a minimum of 3 ft.

(c) Horizontal Clearances. Not less than 3 ft.

(d) Final Spans. Those may be attached to a building but must clear operable windows, fire escapes, and the like by 3 ft. If above the top level of a window then it can be reduced to less than 3 ft. There is additional information concerning openings in commercial buildings.

(e) Zone for Fire Ladders. Allow for fire ladders—at least a 6-ft.-wide space.

225-20 Mechanical Protection of Conductors

Refer to Section 230-50.

225-21 Multiconductor Cables on Exterior Surfaces of Buildings

Refer to Section 230-51.

225-22 Raceways on Exterior Surfaces of Buildings

225-23 Underground Circuits

Refer to Section 300-5.

225-24 Outdoor Lampholders

225-25 Location of Outdoor Lampholders

225-26 Vegetation

ARTICLE 230 SERVICES

230-1 Scope

This article describes requirements for service conductors and equipment for control and protection of services. Diagram 230-1 on page 70-90 of the *NEC*® describes where in Article 230 different parts of the service are described.

A. General

230-2 Number of Services

A building normally can have only one service. If more than one is permitted as a result of the exceptions provided, then a directory of all services must be placed at each service. Reference is made to Section 225-8(d). More than one service is permitted as follows:

(a) For a fire pump if required.

(b) For emergency or standby services.

(c) In multiple-occupancy buildings where there is no room for each tenant to have access to the service, with permission, more than one set of service conductors can be tapped from one service drop or service lateral.

(d) As a result of capacity requirements when the capacity exceeds 2000 A at a voltage not more than 600 V, the utility company requires it as a result of the load, or by special permission.

(e) If the building is very large, with special permission.

(f) When different voltages, frequencies, etc. are required.

(g) When underground conductors are sized at least 1/0 and are connected at the supply but not at the load and service equipment for use only for the purpose of Section 230-40, Exception No. 2.

230-3 One Building or Other Structure Not to Be Supplied through Another

Service conductors cannot pass through another building.

230-6 Conductors Considered Outside of Building

If conductors have a cover of at least 2 in. of concrete under a building or are installed in a raceway encased in at least 2 in. of concrete or are installed in a transformer vault as per Article 450 Part C, they are treated as being outside the building.

230-7 Other Conductors in Raceway or Cable

230-8 Raceway Seal

230-9 Clearance from Building Openings

If installed as open conductors or multiconductor cable without an outer jacket, clearance of at least 3 ft. is required from operable windows, doors, porches, fire escapes, and the like. Service conductors are not permitted underneath openings through which materials can be moved or obstructing these openings.

230-10 Service Conductors

The service conductors are those from the service point to the service disconnecting means.

B. Overhead Service—Drop Conductors

230-21 Overhead Supply

Definition of service drop.

230-22 Insulation or Covering

Insulation must be either thermoplastic or thermosetting and be capable of being exposed to the atmosphere and other conditions without leakage current.

230-23 Size and Rating

Conductors should have the capacity required for the load in accordance with Article 220 and have enough mechanical strength. The minimum size should be No. 8 copper or No. 6 aluminum or copper-clad aluminum. There is an exception.

230-24 Clearances

(a) Eight feet maintained in all directions 3 ft. from the edge of the roof is generally the clearance required above roofs with a nominal voltage of 600 V except where the service drop is attached to the side of the building. This is reduced to 3 ft. with a voltage of 300 V and a roof slope of 4 in. in 12 in. or more. It is also reduced to 18 in. above the overhanging portion with a nominal voltage of 300 V if only up to 4 ft. of service drop conductors pass over the roof and the conductors end on a support or raceway. Where there is traffic above a roof surface use requirements of Section 230-24(b).

(b) The clearances above the ground vary for conductors with a nominal voltage of 600 V. They are:

10 ft.: voltage of 150 V to ground, service entrance to buildings accessible to pedestrians.
12 ft. over residential property and no truck traffic commercial areas with voltage to ground not in excess of 300.
15 ft.: at areas listed in 12 ft. classification when the voltage exceeds 300 V to ground.
18 ft.: over public streets. Private property is included if traveled with commercial vehicles.

(c) Refer to Section 230-9 for clearances from building openings.

(d) Refer to Section 680-8 for clearances from swimming pools.

230-26 Point of Attachment

A minimum clearance of 10 ft. is required. Compliance with Section 230-24 is required.

230-27 Means of Attachment

230-28 Service Masts as Support

230-29 Supports over Buildings

C. Underground Service—Lateral Conductors

230-30 Insulation

Generally, the voltage used determines the voltage rating of the insulation. There are some exceptions for grounded conductors of bare wire and aluminum or copper-clad aluminum cable assemblies.

230-31 Size and Rating

The conductors must be large enough to carry the load in accordance with Article 220. No smaller than No. 8 for copper and No. 6 for aluminum or copper-clad aluminum. An exception is made where the service supplies only a single load on a single branch circuit. The grounded conductor must not be sized less than required in Section 250-23(b).

230-32 Protection against Damage

D. Service-Entrance Conductors

230-40 Number of Service-Entrance Conductor Sets

Generally, a service drop or lateral can only supply one set of service conductors. However, one set of service conductors can supply each occupancy or a group of occupancies in a multiple-occupancy building. A second exception for separate enclosures is noted. A third exception is noted for buildings separated a minimum of 15 ft.

230-41 Insulation of Service-Entrance Conductors

Generally, if the conductors are on or enter a building, they should be insulated. Exceptions are made for grounded conductors of bare copper wire and aluminum or copper-clad aluminum cable assemblies in special cases.

230-42 Size and Rating

Generally, the conductors are required to be large enough to carry the load. Requirements for ungrounded conductors are that a minimum of 100 A three-wire is required for a one-family dwelling with either six or more two-wire branch circuits or a load of at least 10 kVA. A minimum of 60 A is required for other loads. There are three exceptions. The grounded conductor must comply with Section 250-23(b).

230-43 Wiring Methods for 600 Volts, Nominal, or Less

Only the following methods are permitted: Open wiring on insulators; type IGS cable; rigid metal conduit; intermediate metal conduit; EMT; ENT; service entrance cables; wireways; busways; auxiliary gutters; rigid NMC; cablebus; MC cable; mineral-insulated, metal-sheathed cable; flexible metal conduit or liquidtight flexible metal conduit with certain restrictions; or liquidtight flexible nonmetallic conduit. To support cables, a cable tray system may be used.

230-46 Unspliced Conductors

230-49 Protection against Damage—Underground

230-50 Protection of Open Conductors and Cables against Damage—Above Ground

230-51 Mounting Supports

230-52 Individual Conductors Entering Buildings or Other Structures

230-53 Raceways to Drain

230-54 Overhead Service Locations

Connections have to be made raintight with a raintight head or with a gooseneck properly taped and painted or taped with self-sealing, weather-resistant thermoplastic. The heads or goosenecks must be above the point where the conductors are attached to the building and must be securely attached. Drip loops must be constructed to prevent entrance of moisture. There are other requirements.

230-55 Termination at Service Equipment

230-56 Service-Entrance Conductor with the Higher Voltage to Ground

E. Service Equipment—General

230-62 Service Equipment—Enclosed or Guarded

Energized parts must be enclosed or installed in switchboard or similar equipment as stated in Sections 110-17 and 110-18.

230-63 Grounding and Bonding

230-64 Working Space

230-65 Available Short Circuit

230-66 Marking. Service equipment must be marked for use as service equipment

F. Service Equipment—Disconnecting Means

230-70 General

There must be a way to disconnect all conductors in the building from the service-entrance conductors. The disconnect must be readily accessible either outside the building or inside the building nearest the point of entrance of the service conductors. It must have proper markings and be suitable for its use. Service disconnects cannot be installed in bathrooms.

230-71 Maximum Number of Disconnects

The provision for disconnecting each service (Section 230-2) or the service entrance conductors (Section 230-40, Exception No. 1) can consist of no more than six switches or six circuit breakers. Single-pole units can be used for multiwire circuits, one pole for each ungrounded conductor. They must be tied together as a multipole disconnect and will then be counted as one disconnect. An exception is made for the control circuit of a ground-fault protection system.

230-72 Grouping of Disconnects

All disconnects are to be grouped together. Exceptions are for services mentioned in Section 230-2 and when a switch is used for a fire pump. The service disconnects must be located near the point of entry to the building and must be readily accessible. Where multiple occupancies are present, each occupant must have access to his or her own disconnect. An exception is made to this latter requirement for certain management conditions.

230-74 Simultaneous Openings of Poles

230-75 Disconnection of Grounded Conductor

230-76 Manually or Power Operable

230-77 Indicating

230-78 External Operable

The disconnect must be operable from the outside without the operator coming into contact with energized parts.

230-79 Rating of Disconnect

In general, the service disconnect must be sized to carry the calculated load. There are special cases where a minimum size is required. They are:

15 A: a single branch-circuit load
30 A: up to two, two-wire branch circuits
100 A: three-wire: one-family dwelling with six or more two-wire branch circuits or calculated load is 10 kW or more
60 A: all other cases

230-80 Combined Ratings of Disconnects

If more than one service switch is used, their combined rating must be at least the rating shown in Section 230-79.

230-81 Connection to Terminals

230-82 Equipment Connected to the Supply Side of Service Disconnect

Among the pieces of equipment allowed to be connected to the supply side of a service switch are: current-limiting devices, meters, disconnecting devices used as service equipment, meters, instrument transformers, surge protective devices, emergency circuits, fire and sprinkler alarms, fire pumps and interconnected electric production sources. Some of these items require separate protection and disconnects. Refer to the *NEC*®
for a complete description and list.

230-83 Transfer Equipment

G. Service Equipment—Overcurrent Protection

230-90 Where Required

(a) Every ungrounded conductor must have an overcurrent device. It must normally not be set higher than the rating of the conductors. There are five exceptions to this and they include: motor starting current; where no exact size exists; six fuses or circuit breakers is the limit for consideration as a service overcurrent device; fire pumps under certain conditions. When permitted by Note 3 to the Ampacity Tables, 0-2000 volts for 120/240 volt, 3-wire, single-phase dwelling services.

(b) No overcurrent device is to be placed in an ungrounded conductor.

230-91 Location of Overload Protection

Each occupant of a multiple dwelling must have access to his or her own overcurrent protection. The service overcurrent device must be part of the service disconnect or be next to it.

230-92 Locked Service Overcurrent Devices

230-93 Protection of Specific Circuits

230-94 Relative Location of Overcurrent Devices and Other Service Equipment

Some of the items allowed ahead of the overcurrent device are: service switch, surge protectors, emergency supply devices, protective signaling systems, fire pump, meters, and power-operable service equipment controls.

230-95 Ground-Fault Protection of Equipment

This is required for all services that are 150 V to ground up to 600 V phase to phase, solidly grounded wye, with service switches 1000 A or more. The rest of this section contains information about the setting of these devices, the use of switches and fuses, and performance testing. An exception is allowed for continuous industrial processes where a non-orderly shutdown would be hazardous and for fire pumps.

H. Services Exceeding 600 Volts Nominal

230-200 General

230-202 Service-Entrance Conductors

230-203 Warning Signs

230-204 Isolating Switches

230-205 Disconnecting Means

230-206 Overcurrent Devices as Disconnecting Means

230-208 Protection Requirements

230-209 Surge Arresters (Lighting Arresters)

230-210 Service Equipment—General Provisions

230-211 Metal-Enclosed Switchgear

ARTICLE 240
OVERCURRENT PROTECTION

240-1 Scope

Parts A through G give requirements for circuits of up to 600 V. Part H covers over 600 V.

A. General

240-2 Protection of Equipment

Equipment must be protected according to the following articles.

	Article
Air-Conditioning and Refrigerating Equipment	440
Appliances	422
Branch Circuits	210
Busways	364
Capacitors	460
Class 1, Class 2, and Class 3 Remote-Control, Signaling, and Power-Limited Circuits	725
Closed-Loop and Programmed Power Distribution System	780
Cranes and Hoists	610
Electric Signs and Outline Lighting	600

	Article
Electric Welders	630
Electrolytic Cells	668
Elevators, Dumbwaiters, Escalators, Moving Walks, Wheel Chair Lifts, and Stairway Chair Lifts	620
Emergency Systems	700
Fire Protective Signaling Systems	760
Fire Pumps	695
Fixed Electric Heating Equipment for Pipelines and Vessels	427
Fixed Electric Space Heating Equipment	424
Fixed Outdoor Electric De-icing and Snow-Melting Equipment	426
Generators	445
Health Care Facilities	517
Induction and Dielectric Heating Equipment	665
Industrial Machinery	670
Motion Picture and Television Studios and Similar Locations	530
Motors, Motor Circuits, and Controllers	430
Phase Converters	455
Pipe Organs	650
Places of Assembly	518
Services	230
Solar Photovoltaic Systems	690
Sound-Recording and Similar Equipment	640
Switchboards and Panelboards	384
Theaters, Audience Areas of Motion Picture and Television Studios, and Similar Locations	520
Transformers and Transformer Vaults	450
X-Ray Equipment	517 660

240-3 Protection of Conductors

For other than flexible cords and fixture wires, use ampacities specified in Section 310-15 when calculating overcurrent protection. There are thirteen cases where this rule does not necessarily hold true with special requirements: power loss hazard;

devices rated 800 amperes or less; devices rated over 800 amperes; tap conductors; motor operated appliance circuit conductors; motor and motor control circuit conductors; phase converter supply conductors; air conditioning and refrigeration equipment circuit conductors; transformer secondary conductors; capacitor circuit conductors; electric welder circuit conductors; remote control signalling and power limited circuit conductors; and fire protective signalling system circuit conductors.

240-4 Protection of Fixture Wires and Cords

240-6 Standard Ampere Ratings

The standard ratings for fuses and inverse time circuit breakers are: 15, 20, 25, 30, 35, 40, 45, 50, 60, 70, 80, 90, 100, 110, 125, 150, 175, 200, 225, 250, 300, 350, 400, 450, 500, 600, 700, 800, 1000, 1200, 1600, 2000, 2500, 3000, 4000, 5000, and 6000 A. Fuses are also rated at 1, 3, 6, 10, and 601 A. There is a requirement for circuit breakers having external adjustable means.

240-8 Fuses or Circuit Breakers in Parallel

240-9 Thermal Devices

240-10 Supplementary Overcurrent Protection

240-11 Definition of Current-Limiting Overcurrent Protective Device

240-12 Electrical System Coordination

If an orderly shutdown is required to reduce hazards to equipment and personnel, the following method can be used; coordinated short-circuit protection and/or overload indication.

240-13 Ground-Fault Protection of Equipment

For solidly grounded wye systems that are more than 150 V to ground and not exceeding 600 V phase to phase, groundfault protection of equipment must be provided for each main disconnecting means rated 1000 A or more, in accordance with Section 230-95. The three exceptions are where an orderly shutdown in an industrial process is required, for fire pumps, and if the disconnect is protected by service groundfault protection.

B. Location

240-20 Ungrounded Conductors

An overcurrent device (fuse or circuit breaker) has to be connected in series with each ungrounded conductor. An overcurrent relay with a current transformer is allowed. Generally, a circuit breaker has to open all ungrounded conductors. Single-pole circuit breakers are permitted under certain conditions listed in this section. In a closed-loop power distribution system, listed devices can be used instead of fuses or circuit breakers.

240-21 Location in Circuit

This section lists the location of the overcurrent device in relation to the ungrounded circuit conductor.

(a) Feeder and Branch Circuit Conductors. In general, the overcurrent device must be located at the supply to the conductors. Sections (b) through (n) following, indicate exceptions to this rule.

(b) Feeder Taps Not Over 10 Feet Long. They may be tapped to a feeder or transformer secondary without overcurrent protection when the tap is not longer than 10 ft.; the rating of the tap conductor is not less than the loads and devices it supplies and the overcurrent device at the end of the tap; the tap ends at the device it supplies; and the rating of the overcurrent device on the line side of the tap is not more than 1,000 percent of the conductors if they leave the enclosure where the tap is made.

(c) Feeder Taps Not over 25 Feet Long. When the tap is not longer than 25 ft. it does not require protection if the ampacity of the tap is at least one-third the rating of the feeder or overcurrent protection it was tapped from; ends in a single overcurrent device rated for the device; and is mechanically protected.

(d) Feeder Taps Supplying a Transformer (Primary Plus Secondary Not Over 25 Feet Long). The rule is similar to the 25 ft. tap (item c above). No overcurrent protection is required for a tap for a feeder supplying a transformer if the ampacity of the primary wires is at least one-third the rating of the overcurrent device protecting the feeder; if both primary and secondary wires are not longer than 25 ft. combined and both ratings (including transformer ratio for secondary) are not larger than one-third the rating of the circuit being tapped; the tap is mechanically protected; and the secondary conductors end in a single overcurrent device limiting the load current to that permitted by Section 310-15.

(e) Feeder Taps Over 25 Feet Long. No overcurrent protection is required for a tap located in manufacturing installations with bays over 35 ft. in height if the horizontal length is not longer than 25 ft. with the overall length not longer than 100 ft. All the rules of the 25 ft. tap rule (c) apply, and in addition, the tap conductors must be at least No. 6 copper or No. 4 aluminum; they can not go through floors, walls, or ceilings; the tap must be least 30 ft. from the floor; and the tap conductors have no splices and are continuous from one end to the other.

(f) Branch Circuit Taps. The branch circuit overcurrent device, if in accordance with Sections 210-19, 210-20, and 210-24, can be used for a tap to individual outlets and conductors to a single electric household range.

(g) Busway Taps. Busways and their taps can be protected in accordance with Sections 364-10 through 364-14.

(h) Motor Circuit Taps. Motor branch circuit conductors can be protected as indicated in Sections 430-28 and 430-53.

(i) Conductors from Generator Terminals. They can be protected as indicated in Section 445-5.

(j) Transformer Secondary Conductors of Separately Derived Systems for Individual Installations. This is discussed.

(m) Outside Feeder Taps. These are discussed.

(n) Service Conductors. They can be protected as indicated in Section 230-91.

240-22 Grounded Conductors

No overcurrent device should be placed in series with a grounded conductor unless it simultaneously opens all other conductors of the circuit or when required by Sections 430-36 and 430-37.

240-23 Change in Size of Grounded Conductor

240-24 Location in or on Premises

Overcurrent devices must be readily accessible and each occupant must have access to the ones protecting the conductors for his or her occupancy. The overcurrent devices must not be exposed to physical damage and must not be located near easily ignitable material or in bathrooms of dwelling units or guest rooms of hotels and motels. Refer to the *NEC*® for certain exceptions.

C. Enclosures

240-30 General

240-32 Damp or Wet Locations

240-33 Vertical Position

D. Disconnecting and Guarding

240-40 Disconnecting Means for Fuses

In general, the disconnection means must be provided on the supply side of the fuse. This does not hold true for service current-limiting devices as per Section 230-82 or single disconnecting devices for multiple sets of fuses as per Section 430-112 (generators) and Section 424-22 (fixed electric space heating).

240-41 Arcing or Suddenly Moving Parts

E. Plug Fuses, Fuseholders, and Adapters

240-50 General

These are not permitted in circuits where the voltage between conductors is more than 125 V except with a grounded neutral and not more than 150 V to ground. There are other requirements.

240-51 Edison-Base Fuses

240-52 Edison-Base Fuseholders

240-53 Type S Fuses

240-54 Type S Fuses, Adapters, and Fuseholders

F. Cartridge Fuses and Fuseholders

240-60 General

These are not permitted in circuits operating at over 300 V between conductors if rated 300 V. There is an exception for single phase line to neutral circuits where this voltage is not larger than 300 V and the circuit is supplied from a three-phase, four-wire, solidly grounded neutral system. The holders must prevent interchangeability where a fuse can be put into a circuit of lower ampere rating or higher voltage rating. All fuses must be marked, indicating the current, voltage, and interrupting rating, if it is current limiting, and the manufacturer's name and trademark. The interrupting rating is not required when they are used for supplementary protection.

240-61 Classification

G. Circuit Breakers

240-80 Method of Operation

Normally, they must be trip free and capable of manual operation in addition to other normal means, such as electrical or pneumatic operation.

240-81 Indicating

Must indicate on or off—on is up position if mounted vertically.

240-82 Nontamperable

240-83 Marking

They must all be properly marked. When properly used as a switch, they must be approved for switching duty and marked SWD. There are requirements for voltage marking, interrupting ratings, and location.

240-85 Applications

H. Overcurrent Protection Over 600 Volts, Nominal

240-100 Feeders

240-101 Branch Circuits

ARTICLE 250 GROUNDING

A. General

250-1 Scope

Generally gives requirements for grounding and bonding.

250-2 Application of Other Articles

The following other articles have special grounding requirements:

	Article	Section
Agricultural Buildings		547-8
Appliances		422-16
Branch Circuits		210-5
		210-6
		210-7
Cablebus		365-9
Capacitors		460-10
		460-27
Circuits and Equipment Operating at Less Than 50 Volts	720	
Class 1, Class 2, and Class 3 Circuits Remote Control, Signaling, and Power Limited Circuits		725-20 725-43
Closed-Loop and Programmed Power Distribution		780-3
Communications Circuits	800	
Community Antenna Television and Radio Distribution Systems		820-33 820-40 820-41
Conductors for General Wiring	310	
Cranes and Hoists	610	
Electrically Driven or Controlled Irrigation Machines		675-11(c) 675-12 675-13 675-14 675-15
Electric Signs and Outline Lighting	600	
Electrolytic Cells	668	
Electronic Computer/Data Processing Equipment		645-15

	Article	Section
Elevators, Dumbwaiters, Escalators, Moving Walks, Wheel Chair Lifts, and Stairway Chair Lifts	620	
Fire Protective Signaling Systems		760-6
Fixed Electric Heating Equipment for Pipelines and Vessels		427-21 427-29 427-48
Fixed Electric Space Heating Equipment		424-14
Fixed Outdoor Electric De-icing and Snow-Melting Equipment		426-27
Fixtures and Lighting Equipment		410-17 410-18 410-19 410-21 410-105(b)
Flexible Cords		400-22 400-23
Floating Buildings		553-8 553-10 553-11
Grounding-Type Receptacles, Adapters, Cord-Connectors, and Attachment Plugs		410-58
Hazardous (Classified) Locations	500-517	
Health Care Facilities	517	
Induction and Dielectric Heating Equipment	665	
Industrial Machinery	670	
Intrinsically Safe Systems	504-50	
Lighting Fixtures, Lampholders, Lamps, and Receptacles	410	
Marinas and Boatyards		555-7
Mobile Homes and Mobile Home Parks	550	
Motion Picture and Television Studios and Similar Locations		530-20 530-66

	Article	Section
Motors, Motor Circuits, and Controllers	430	
Outlet, Device, Pull and Junction Boxes, Conduit Bodies, and Fittings		370-4 370-25
Over 600 Volts, General		710-4(b)(1)
Panelboards		384-20
Pipe Organs	650	
Radio and Television Equipment	810	
Receptacles and Cord Connectors		210-7
Recreational Vehicles and Recreational Vehicle Parks	551	
Services	230	
Service Equipment		230-63
Solar Photovoltaic Systems		690-41 690-42 690-43 690-44
Sound-Recording and Similar Equipment		640-4
Swimming Pools, Fountains, and Similar Installations	680	
Switchboards and Panelboards		384-3(d) 384-11
Switches		380-12
Theaters, Audience Areas of Motion Picture and Television Studios, and Similar Locations		520-81
Transformers and Transformer Vaults		450-10
Use of Identification of Grounded Conductors	200	
X-Ray Equipment	660	517-151

B. Circuit and System Grounding

250-3 Direct-Current Systems

250-5 Alternating-Current Circuits and Systems to Be Grounded

The following circuits and systems must be grounded. Others may be grounded if desired.

(a) Alternating-current circuits of less than 50 V must be grounded if supplied by transformer systems of more than 150 V to ground or by ungrounded transformer systems. All outdoor overhead circuits must be grounded.

(b) Alternating-current circuits between 50 and 1000 V of the following types must be grounded:

1. Systems of up to 150 V to ground from an ungrounded conductor
2. Systems that are three-phase, four-wire with a neutral circuit conductor
3. Three-phase four-wire delta systems with the midpoint of one phase used as a circuit conductor
4. Uninsulated service conductor as per Section 230-22, 230-30, and 230-41

There are a number of exceptions and the reader is referred to Section 250-5(b), Exception Nos. 1 through 5.

(c) Alternating-Current Systems 1 kV and Over

(d) Separately Derived Systems. If the source is separately derived, has no connection to another source, or is grounded and required to be grounded, it must be grounded as required by Section 250-26.

250-6 Portable and Vehicle-Mounted Generators

250-7 Circuits Not to Be Grounded

Those specified in Article 517, electric cranes used in areas of Class III over combustible fibers as indicated in Section 503-13, and Electrolytic Cells noted in Article 668.

C. Location of System Grounding Connections

250-21 Objectionable Current over Grounding Conductors

250-22 Point of Connection for Direct Current Systems

250-23 Grounding Service-Supplied Alternating-Current Systems

As required, each service must have a grounding electrode conductor connected to a grounding electrode in compliance with Part H of Article 250. The reader should be familiar with the requirements of this article in the *NEC*® as there are a number of exceptions.

250-24 Two or More Buildings or Structures Supplied from a Common Service

Each building must have a grounding electrode (refer to Part H) connected to the building disconnecting devices metal enclosure. In a grounded system, the electrode must also connect to the grounded conductor of the ac system on the supply side of the building disconnecting device. There are exceptions to this. There are requirements for ungrounded systems and disconnecting means located in a separate building or structure on the same premises.

250-25 Conductor to Be Grounded—Alternating-Current Systems

One conductor must be grounded in a two-wire single-phase system. The neutral conductor is grounded in a three-wire single-phase system and a multiphase system where one phase is used as a three-wire single-phase system. The common conductor is grounded in a multiphase system with one wire common to all phases. A phase conductor is grounded in a multiphase system requiring it.

250-26 Grounding Separately Derived Alternating-Current Systems

250-27 High-Impedance Grounded Neutral System Connections

D. Enclosure and Raceway Grounding

250-32 Service Raceways and Enclosures

These metal enclosures and raceways must be grounded. There is an exception for a metal elbow installed in an underground installation under certain conditions.

250-33 Other Conductor Enclosures and Raceways

These metal enclosures and raceways must be grounded. Note four exceptions for open wire: knob and tubing and nonmetallic-sheathed cable, enclosures used to physically protect cable assemblies, enclosures not required to be grounded [Section 250-43(i)], and metal elbows under certain conditions.

E. Equipment Grounding

250-42 Equipment Fastened in Place or Connected by Permanent Wiring Methods (Fixed)

Metal enclosures must be grounded if located in a wet or damp location and not isolated, if they can be touched by people and are 8 ft or less vertically and 5 ft or less horizontally to ground or a grounded object, if in contact with metal, if in a hazardous location, if metal-clad type of wiring is used under certain conditions, and if the voltage is greater than 150 V to ground. There are four exceptions to the last case.

250-43 Fastened in Place or Connected by Permanent Wiring Methods (Fixed)—Specific

All exposed metal parts that do not conduct current of the following kinds of equipment and enclosures must be grounded: switchboard frames and structures except two-wire dc where insulated from ground: pipe organs; motor frames; enclosures for motor controllers; elevators and cranes; all electric equipment in commercial garages, theaters, and motion picture studios except pendant lampholders of 150 V or less: electric signs, outline lighting and associated equipment as noted in Article 600; motion picture equipment; power limited remote-control, signaling, and fire protective signaling equipment supplied by Class 1, 2, and 3 circuits where required; lighting fixtures as per Article 410; motor operated water pumps, and metal well casings. There are some exceptions.

250-44 Nonelectric Equipment

The metal parts of this equipment must be grounded: cranes and hoists; elevator cars of nonelectric type; electric elevators, metal partitions enclosing conductors of 1 kV and over to ground; mobile homes and recreational vehicles per Articles 550 and 551.

250-45 Equipment Connected by Cord and Plug

There are areas where exposed metal parts (noncurrent-carrying) must be grounded even though connected by cord and plug. This is true of hazardous locations and circuits over 150 V to ground; three exceptions are motors when guarded, electrically heated appliances with special exemptions and insulation to ground and listed equipment double-insulated under certain conditions. In a residence the following items are required to have equipment grounding: refrigerators, freezers, and air conditioners; clothes washing, clothes drying equipment, dishwashing machines, kitchen waste disposers, sump pumps and electrical aquarium equipment; hand-held motor-operated tools, stationary and fixed motor operated tools, light industrial motor operated tools; motor-operated hedge clippers, lawn mowers, snow blowers, wet scrubbers, etc.: and portable handlamps. If the appliance has double insulation and is so listed, equipment grounding is not necessary. In other than residences, the list is the same as residences, with the addition of cord- and plug-connected appliances in damp or wet locations and portable headlamps, electronic computer data processing equipment, tools used in wet or conductive locations. The same exception as in residences is in effect, with an additional exception for tools in wet or conductive locations if supplied through an isolating transformer with a secondary ungrounded and not over 50 V.

250-46 Spacing from Lightning Rods

F. Methods of Grounding

250-50 Equipment Grounding Conductor Connectors

250-51 Effective Grounding Path

250-53 Grounding Path to Grounding Electrode at Services

A grounded item (equipment grounding conductors, service equipment enclosures, and the grounded service conductors) must be connected to the grounding electrode. There is an exception for a high-impedance grounded neutral system. A requirement is noted for the main bonding jumper with two exceptions.

250-54 Common Grounding Electrode

Where the ac system is grounded, the system and equipment must be grounded to the same electrode or electrodes bonded together

250-55 Underground Service Cable

The metal sheath of an underground cable system or its metal conduit does not have to be grounded at the building, assuming that it is a continuous system.

250-56 Short Sections of Raceway

250-57 Equipment Fastened in Place or Connected by Permanent Wiring Methods (Fixed)—Grounding

250-58 Equipment Considered Effectively Grounded

(a) This is true when it is electrically connected to a supporting metal rack or structure that is properly grounded. The building's structural steel is not considered an effective grounding method.

(b) Metal car frames

250-59 Cord- and Plug-Connected Equipment

250-60 Frames of Ranges and Clothes Dryers

250-61 Use of Grounded Circuit Conductor for Grounding Equipment

(a) Generally, a grounded circuit conductor can be used to ground equipment on the supply side of the service disconnect, supply side of main disconnects for a separate building (Section 250-24), and in a separate, derived system the supply side of the disconnecting or overcurrent devices.

(b) Generally, a grounded circuit conductor cannot be used to ground equipment on the load side of the service disconnect. There are five exceptions to this:

1. As indicated in Section 250-60 for ranges, wall-mounted ovens, etc.
2. For separate buildings as indicated in Section 250-24.

3. Grounding meter enclosures if no service ground-fault protection is used, the meter is near the service switch, and the grounded circuit conductor is not smaller than noted in Table 250-95 for equipment grounding conductors.
4. In accordance with Sections 710-72(e)(1) and 710-74.
5. Dc systems in accordance with Section 250-22, Exception.

250-62 Multiple Circuit Connections

G. Bonding

250-70 General

Bonding must be provided to make sure there is proper electrical continuity in the grounding system.

250-71 Service Equipment

All service equipment must be bonded, items are listed. A means must be provided for intersystem bonding.

250-72 Methods of Bonding Service Equipment

There are five approved methods:

(a) For the grounded service conductor use a method noted in Section 250-113.

(b) Threaded couplings for rigid metal conduit and intermediate metal conduit

(c) Threadless couplings and connectors for rigid and intermediate metal conduit and electrical metallic tubing

(d) Bonding jumpers around knockouts

(e) Other devices

250-73 Metal Armor or Tape of Service Cable

250-74 Connecting Receptacle Grounding Terminal to Box

Generally, a bonding jumper must be used. There are four exceptions:

1. If the box is service mounted and the yoke makes good electrical contact with the box. There are restrictions on cover-mounted receptacles.
2. Where the receptacle is listed as such with special yokes and screws.

3. Floor boxes so listed.

4. Where there is a requirement for the reduction of electrical noise, an insulated grounding conductor can be used. It is run with the circuit conductors. It can pass through panelboards without connection to the grounding terminal. The grounding terminal in the receptacle can be insulated from the mounting.

250-75 Bonding Other Enclosures

All items that will be used as grounding conductors must be bonded. There is an exception.

250-76 Bonding for Over 250 Volts

250-77 Bonding Loosely Jointed Metal Raceways

250-78 Bonding in Hazardous (Classified) Locations

All items in a hazardous (classified) location must be bonded as required by Section 250-72, regardless of the voltage.

250-79 Main and Equipment Bonding Jumpers

250-80 Bonding of Piping Systems and Exposed Structural Steel

H. Grounding Electrode System

250-81 Grounding Electrode System

The following four items and any electrodes as per Section 250-83(c) and (d) if they exist must be bonded together to form the grounding electrode system. Any one of them can be considered the grounding electrode. The grounding electrode conductor must be unspliced and sized for the largest grounding electrode conductor required by the electrodes. There is an exception. Interior metal water piping must be located within 5 ft. from the point of entrance. An exception is made for industrial and commercial locations.

(a) Metal Underground Water Pipe. This must be underground for at least 10 ft. and an additional electrode is required.

(b) Metal Frame of the Building Effectively Grounded

(c) Concrete—Encased Electrode. The electrode must

be at least 20 ft. of No. 4 wire or 20 ft. of steel reinforcing rods at least ½ in. in diameter. It must be encased in a minimum of 2 in. of concrete near the bottom of the foundation.

(d) Ground Ring

250-83 Made and Other Electrodes

Other methods may be used when none of the ones in Section 250-81 are available. These are local metal underground systems or structures, rod and pipe electrodes, and plate electrodes. The reader is referred to this section in the *NEC*® for complete details. The metal underground gas piping system and aluminum electrodes cannot be used.

250-84 Resistance of Made Electrodes

250-86 Use of Lightning Rods

They are not to be used for grounding electrodes.

J. Grounding Conductors

250-91 Material

(a) Grounding Electrode Conductor. The material is to be solid or stranded copper, aluminum, or copper-clad aluminum. It can be bare, covered, or insulated and must not contain a splice. There are three exceptions where splices are permitted.

1. In busbars.
2. Taps are permitted where there is more than one service enclosure.
3. The exothermic welding process can be used to extend the grounding electrode conductor or irreversible compression type connectors.

(b) Types of Equipment Grounding Conductors. The following may be used for equipment grounding conductors.

1. Copper or other corrosion-resistant conductor. As with the grounding electrode, it can be solid or stranded, and bare, covered, or insulated.
2. Rigid metal conduit.
3. Intermediate metal conduit.
4. Electrical metallic tubing.

5. Flexible metal conduit (approved for grounding).
6. Armor of Type AC cable.
7. The copper sheath of MI metal sheathed cable.
8. The sheath of MC cable.
9. Cable trays.
10. Cablebus framework permitted in Section 365-2(a).
11. Other metal raceways approved for this purpose. Flexible metal conduit and tubing are allowed if the length is not more than 6 ft, the circuit protection does not exceed 20 A, and terminations are approved for grounding. Where conductors have overcurrent devices more than 20 A and not more than 60 A listed liquidtight flexible metal conduit between ¾ in. and 1¼ in. can be permitted if it is not longer than 6 ft., trade sizes of ⅜ in. and ½ in. are not used in the ground path, and terminations are listed. The length restrictions are for a total of the combined length of liquidtight flexible metal conduit, flexible metal conduit and flexible metallic tubing in the same ground return path.

(c) Supplementary grounding conductors may be used.

250-92 Installation

(a) Grounding Electrode Conductor. If it is No. 4 copper or aluminum or larger, it must be protected if it is subject to physical damage that is severe. Number 6 wires, if fastened to the building and free from physical damage, do not need protection; if not, they must be protected in the same manner as wires smaller than No. 6. Wires smaller than No. 6 must be enclosed in either rigid metal conduit, intermediate metal conduit, rigid nonmetallic conduit, electrical metallic tubing, or cable armor. In all cases above, the wire or enclosure must be fastened to the surface it is running on. Insulated or bare aluminum or copper-clad aluminum cannot be used in corrosive areas.

(b) Enclosures for Grounding Electrode Conductors. A complete electrical path must be ensured for metal enclosures of grounding electrode conductors. If a raceway is used as protection for a grounding conductor, compliance is required with the appropriate article concerning the raceway.

(c) Equipment Grounding Conductor. If it is a metal enclosure for the conductor, or the conductor, the applicable provisions of the *NEC*® must be followed. An exception is made for existing installations where no equipment grounding conductor is available for branch-circuit extensions.

250-93 Size of Direct-Current Grounding Electrode Conductor

It cannot be smaller than the neutral conductor for a three-wire balancer set or balancer winding. It cannot be smaller than the largest conductor if other than as stated in the preceding sentence. It can never be smaller than No. 8 copper or No. 6 aluminum. There are a number of exceptions.

250-94 Size of Alternating-Current Grounding Electrode Conductor

These minimum sizes are given in Table 250-94. There is an exception.

250-95 Size of Equipment Grounding Conductors

These minimum sizes are given in Table 250-95. When circuit conductors are run in parallel, the equipment grounding conductor must be run in parallel and sized according to the overcurrent device rating shown in the table. The conductor never has to be larger than the overcurrent device protecting the circuit conductor to the equipment. There are other requirements and exceptions.

250-97 Outline Lighting

250-99 Equipment Grounding Conductor Continuity

K. Grounding Conductor Connections

250-112 To Grounding Electrode

It must be accessible and provide an effective and permanent ground.

250-113 To Conductors and Equipment

Use only listed connectors and not ones that depend on solder.

250-114 Continuity and Attachment of Equipment Grounding Conductors to Boxes

250-115 Connection to Electrodes

250-117 Protection of Attachment

250-118 Clean Surfaces

TABLE 250-94 Grounding Electrode Conductor for AC Systems

Size of Largest Service-Entrance Conductor or Equivalent Area for Parallel Conductors		*Size of Grounding Electrode Conductor*	
Copper	*Aluminum or Copper-Clad Aluminum*	*Copper*	*Aluminum or Copper-Clad Aluminum**
2 or smaller	1/0 or smaller	8	6
1 or 1/0	2/0 or 3/0	6	4
2/0 or 3/0	4/0 or 250 kcmil	4	2
Over 3/0 through 350 kcmil	Over 250 kcmil through 500 kcmil	2	1/0
Over 350 kcmil through 600 kcmil	Over 500 kcmil through 900 kcmil	1/0	3/0
Over 600 kcmil through 1100 kcmil	Over 900 kcmil through 1750 kcmil	2/0	4/0
Over 1100 kcmil	Over 1750 kcmil	3/0	250 kcmil

Where multiple sets of service-entrance conductors are used as permitted in Section 230-40, Exception No. 2, the equivalent size of the largest service-entrance conductor shall be determined by the largest sum of the areas of the corresponding conductors of each set. Where there are no service-entrance conductors, the grounding electrode conductor size shall be determined by the equivalent size of the largest service-entrance conductor required for the load to be served.

* See installation restrictions in Section 250-92(a).

(FPN) See Section 250-23(b) for size of alternating current system grounded conductor brought to service equipment.

TABLE 250-95 Minimum Size Equipment Grounding Conductors for Grounding Raceway and Equipment

Rating or Setting of Automatic Overcurrent Device in Circuit Ahead of Equipment, Conduit, etc., Not Exceeding (Amperes)	*Size*	
	Copper Wire No.	*Aluminum or Copper-Clad Aluminum Wire No.**
15	14	12
20	12	10
30	10	8
40	10	8
60	10	8
100	8	6
200	6	4
300	4	2
400	3	1
500	2	1/0
600	1	2/0
800	1/0	3/0
1000	2/0	4/0
1200	3/0	250 kcmil
1600	4/0	350 kcmil
2000	250 kcmil	400 kcmil
2500	350 kcmil	600 kcmil
3000	400 kcmil	600 kcmil
4000	500 kcmil	800 kcmil
5000	700 kcmil	1200 kcmil
6000	800 kcmil	1200 kcmil

* See installation restrictions in Section 250-92(a).

250-119 Identification of Wiring Device Terminals

L. Instrument Transformers, Relays, etc.

250-121 Instrument Transformer Circuits

250-122 Instrument Transformer Cases

250-123 Cases of Instruments, Meters, and Relays—Operating at Less Than 1000 Volts

250-124 Cases of Instruments, Meters and Relays—Operating Voltage 1 kV and Over

250-125 Instrument Grounding Conductor

M. Grounding of Systems and Circuits of 1 kV and Over (High Voltage)

250-150 General

If these systems are grounded, they must follow all previous sections as well as this section.

250-151 Derived Neutral Systems

250-152 Solidly Grounded Neutral Systems

250-153 Impedance Grounded Neutral Systems

250-154 Grounding of Systems Supplying Portable or Mobile Equipment

250-155 Grounding of Equipment

ARTICLE 280 SURGE ARRESTERS

A. General

280-1 Scope

280-2 Definition

280-3 Number Required

280-4 Surge-Arrester Selection

B. Installation

280-11 Location

280-12 Routing of Surge-Arrester Connections

C. Connecting Surge Arresters

280-21 Installed at Services of Less Than 1000 Volts

280-22 Installed on the Load Side of Services of Less Than 1000 Volts

280-23 Circuits of 1 kV and Over—Surge-Arrester Conductors

280-24 Circuits of 1 kV and Over—Interconnections

280-25 Grounding

3

WIRING METHODS AND MATERIALS

ARTICLE 300
WIRING METHODS

A. General Requirements

300-1 Scope

This article applies to all wiring installations except as permitted in Article 504; Class 1, 2, and 3 circuits referenced in Article 725; fire alarm signaling circuits referenced in Article 760; optical fiber cables referenced in Article 770; communications systems referenced in Article 800; radio and television equipment referenced in Article 810; and community antenna television and radio distribution systems referenced in Article 820. It does not apply to integral parts of equipment (e.g., motors, motor control centers, etc.)

300-2 Limitations

This chapter applies to voltages up to and including 600 V when not otherwise noted and over 600 V only when specifically mentioned in the *National Electrical Code*®.

300-3 Conductors

(a) Single Conductors

(b) Conductors of the Same Circuit. All conductors of the same circuit including the grounded conductor and equipment grounding conductors must be installed in the same raceway, cable, etc. There are exceptions.

(c) Conductors of Different Systems

(1) 600 Volts, Nominal, or Less. Conductors of 600 V or less can occupy the same wiring enclosure if all conductors have insulation rated for the maximum voltage used. Refer to Section 690-4(b) for solar photovoltaic systems and to Section 725-38(a)(2) for Class 2 and Class 3 conductors.

(2) Over 600 Volts, Nominal. Conductors of over 600 V cannot be in the same wiring enclosure with those rated 600 V or less except for secondary wiring to electric discharge lamps, primary leads of electric discharge lamp ballasts, and excitation, control, relay, and ammeter conductors.

Reference should be made to this section in the *NEC*® for a complete explanation.

300-4 Protection against Physical Damage

300-5 Underground Installations

(a) Minimum Cover Requirements. The cover requirements for direct buried cable or conduit are shown in Table 300-5. The reader should refer to it.

(b) Grounding

(c) Underground Cables under Buildings

(d) Protection from Damage

(e) Splices and Taps

(f) Backfill

(g) Raceway Seals

(h) Bushing

(i) Conductors of the Same Circuit

(j) Ground Movement

300-6 Protection against Corrosion

Generally, all material used must be intended for use in the installed environment.

(a) General. Ferrous material must be protected both inside and out by a coating of corrosion-resistant material (e.g., zinc, cadmium, or enamel). Enamel cannot be used outdoors or in wet locations. A specific exception is made for threads at joints.

(b) In Concrete or in Direct Contact with the Earth. If the material is made for this purpose, it can be used.

(c) Indoor Wet Locations. The cable must be installed with at least ¼ in. of air space from the wall or supporting surface when installed exposed. There is an exception.

300-7 Raceways Exposed to Different Temperatures

300-8 Installations of Conductors with Other Systems

300-9 Grounding Metal Enclosures

300-10 Electrical Continuity of Metal Raceways and Enclosures

There must be electrical continuity between all metal raceways, cable armor, and metal enclosures. There are three exceptions.

300-11 Securing and Supporting

300-12 Mechanical Continuity—Raceways and Cables

300-13 Mechanical and Electrical Continuity—Conductors

Splices or taps are not permitted within a raceway. Exceptions to this are noted in Sections 300-15(a), Exception No. 1, 352-7, 352-29, 354-6 Exception, 362-7, 362-21 or 364-8(a).

300-14 Length of Free Conductors at Outlets, Junctions, and Switch Points

At least 6 in. of conductors must be left at every outlet, junction, or switch point unless the conductors are not spliced or terminated at that point.

300-15 Boxes, Conduit Bodies or Fittings—Where Required

(a) Box or Conduit Body. A box or fitting must be used at each conductor splice connection point, outlet, switch point, junction point, or pull point for conduit or raceway connection. Two exceptions are noted. The first is in certain raceways with removable covers. The second is where a fixture is used as a raceway permitted in Section 410-31.

(b) Box Only. A box is generally required where conductors or cable is spliced, connected, pulled, and at utilization devices except as follows:

1. For insulated outlet devices supplied by nonmetallic-sheathed cable, refer to Section 336-11.
2. When using MI cable, metal-sheathed cable with accessible fittings in straight-through splices.
3. Where cable enters or exits from a raceway used for support or physical damage protection. Fittings must be installed to protect the wire from wear.
4. When using nonmetallic-sheathed cable with a box that is secured to a structural member under certain conditions.
5. When a manufactured metallic system is used.
6. When complying with Sections 370-16(c) and 370-28 a conduit body can be used instead of a box.
7. When a device is listed for use without a box in a closed-loop power distribution system.
8. A fitting that is identified for use can be used instead of a box when no conductors are spliced and it is always accessible.
9. For splices and taps in buried cables and conductors as noted in Section 305(e).

(c) Fittings and Connectors. Fittings and connectors can be used for methods for which they are listed and designed.

(d) Equipment. An integral junction box or wiring compartment is permitted if it is part of listed equipment.

300-16 Raceway or Cable to Open or Concealed Wiring

300-17 Number and Size of Conductors in Raceway

In general, the number and size of conductors in a raceway must allow the dissipation of heat, easy installation, and withdrawal.

300-18 Raceway Installations

Raceways must be installed complete between outlet, junction, or splicing points before the conductors are installed. Three exceptions are: (1) for prewired assemblies as per Articles 349 and 350, (2) when necessary for the installation of utilization equipment, and (3) exposed raceways with removable covers.

300-19 Supporting Conductors in Vertical Raceways

Conductors must be supported at the top of vertical risers with additional support as noted in Table 300-19(a) of the *NEC*® when the rise is greater than the values in that table.

300-20 Induced Currents in Metal Enclosures or Metal Raceways

The effects of induced currents in metal enclosures must be minimized. This is done by grouping all phases, neutral, and grounding conductors in one raceway. If individual conductors are used, a slot can be cut between the holes made in the metallic enclosure through which the individual conductor passes. There are some exceptions.

300-21 Spread of Fire or Products of Combustion

300-22 Wiring in Ducts, Plenums, and Other Air-Handling Spaces

(a) Ducts for Dust, Loose Stock, or Vapor Removal. Wiring of any type is prohibited in these types of installations. This includes ventilation of commercial-type cooking equipment.

(b) Ducts or Plenums Used for Environmental Air. The following can be used when the duct or plenum was specifically made for this purpose: Type MI metal-sheathed cable, Type MC cable, EMT flexible metallic tubing, intermediate metal conduit, or rigid metal conduit. When necessary to connect physically adjustable items, regular or liquidtight flexible metal conduit can be used, but only in lengths not longer than 4 ft. Only devices that act on or sense the contained air, such as thermostats, are permitted. If lighting fixtures are installed for maintenance or repair, they must be of the enclosed gasketed type. This paragraph applies primarily to fabricated ducts and plenums.

(c) Other Space Used for Environmental Air. This paragraph applies to spaces other than those in Sections 300-22(a) and (b) for environmental air handling. Only totally enclosed nonventilated insulated busways that do not have a provision for plug-in connections and Type MI cable, Type MC cable without an overall nonmetallic covering, as well as Type AC cable or other factory-assembled multiconductor control or power cable made to be installed in these areas can be installed in this space. Other types of cables and conductors can be used but only when installed in EMT, flexible metallic tubing, intermediate metal conduit, rigid and flexible metal conduit, where accessible surface metal raceway or metal wireway with metal covers or solid bottom metal cable tray with solid metal covers. Also permitted is electric equipment with a metal enclosure or a listed nonmetallic enclosure with low smoke- and fire-resistive qualities suitable for the ambient temperature unless prohibited in another place in the Code. There are five exceptions to the above:

1. Liquidtight flexible metal conduit can be used in lengths not longer than 6 ft.
2. When identified for this use, integral fan systems.
3. Habitable rooms or areas are not included as applicable spaces.
4. Prefabricated cable assemblies of metallic wiring systems without nonmetallic sheathing when approved for these areas.
5. Joist or stud spaces in dwelling units if wiring or equipment passes through perpendicular to long dimension.

(d) Data Processing Systems. Refer to Article 645.

300-23 Panels Designed to Allow Access

B. Requirements for Over 600 Volts, Nominal

300-31 Covers Required

300-32 Conductors of Different Systems

Conductors of systems 600 V or less and systems over 600 V cannot be in the same enclosure except in motors, switchgear, control assemblies, and the like, or in manholes with proper separation.

300-34 Conductor Bending Radius

300-35 Protection against Induction Heating

300-36 Grounding

300-37 Underground Installations

ARTICLE 305
TEMPORARY WIRING

305-1 Scope

305-2 All Wiring Installations

305-3 Time Constraints

305-4 General

305-5 Grounding

305-6 Ground-Fault Protection for Personnel

305-7 Guarding

ARTICLE 310
CONDUCTORS FOR GENERAL WIRING

310-1 Scope

Article 310 is to be used for general requirements for wiring, but not in areas where it is part of an integral device such as a motor, motor controller, or where covered in another part of the *NEC*®.

310-2 Conductors

Conductors must be insulated and made of copper, copper-clad aluminum, or aluminum unless otherwise specified in the *NEC*®.

310-3 Stranded Conductors

Any conductor installed in a raceway must be stranded if it is No. 8 or larger. This is not required if it is permitted or required in another part of the *NEC*®.

310-4 Conductors in Parallel

Conductors 1/0 or larger can be run in parallel if they are the same length, the same size, and of the same material with the same insulation and terminated the same way. If run separately, the raceways must be the same. There are four exceptions to this:

1. Section 620-12(a)(1), Exception.
2. If smaller than No. 1/0, used for control power to indicating instruments, relays, and similar devices, and in the same raceway, each individual conductor can carry the entire load and the overcurrent rating is not larger than the ampacity of any single conductor.
3. Sizes smaller than No. 1/0 are allowed for frequencies of 360 Hz and higher provided that Exception 2 is met.
4. With engineering supervision No. 2 AWG or larger grounded neutral conductors can be run parallel in existing installations.

310-5 Minimum Size of Conductors

Minimum sizes are shown in Table 310-5. For voltages up to 2000, it is No. 14 for copper and No. 12 for aluminum or copper-clad aluminum. The following exceptions apply:

1. Refer to Section 400-12 for flexible cords.
2. No. 18 may be used for fixture wire (Section 410-24).
3. Refer to Section 430-22 for one horsepower or less motors.
4. Refer to Section 610-14 for cranes and hoists.
5. Refer to Section 620-12 for elevator control and signaling circuits.
6. Refer to Sections 725-27 and 725-51 for Class 1, 2, and 3 circuits.
7. Refer to Sections 760-27, 760-51, and 760-71 for fire alarm signaling circuits.
8. Refer to Section 430-72 for motor control circuits.

310-6 Shielding

310-7 Direct Burial Conductors

310-8 Wet Locations

Use only lead-covered conductors; Type RHW, TW, THW, THHW, THWN, or XHHW; or a type listed as such for a wet location.

310-9 Corrosive Conditions

310-10 Temperature Limitations of Conductors

310-11 Marking

All conductors and cables must have marked on them the rated voltage, the proper letter type, the manufacturer's name or trademark, and the size or circular mil area. The size markings should be on the surface and spaced not more than 24 in. for eight types of cables. Other markings may be spaced 40 in. apart. Certain cables can use marker tape inside the cable. Refer to this section in the *NEC*® for specific types where marker tape is permitted, which cables can have the size located elsewhere, and suffix designations for number of conductors. Special characteristics can be surfaced marked on conductors listed in Table 310-13 and Table 310-61.

310-12 Conductor Identification

(a) Grounded Conductors. Insulated grounded conductors generally must have an outer covering of white or natural gray. There are exceptions for fixture wires, Type MI cables, requirements of Section 210-5(a), and where qualified persons only are used for maintenance and supervision as to where markings are permitted.

(b) Equipment Grounding Conductors. When insulated, they must generally have a cover of green or green with yellow stripes. There are exceptions for conductors larger than No. 6 and larger wires as to where the markings are permitted and the same exceptions for maintenance and supervision by qualified persons as to where the markings are permitted.

(c) Ungrounded Conductors. These may be any color except white, natural gray, or green.

310-13 Conductor Constructions and Applications

Insulated conductors must be in accordance with Tables 310-13 and 310-61 through 310-64.

310-14 Aluminum Conductor Material

310-15 Ampacity*

Ampacities for conductors rated 0 through 2000 V can be determined either through use of the tables or, under engineering supervision, the use of the formula noted below. The ratings for

*Note: Section 310-15 continues on page 87.

TABLE 310-16 Allowable Ampacities of Insulated Conductors, Rated 0-2000 Volts, 60° to 90°C (140° to 194°F) Not More Than Three Current Conductors in Raceway or Cable or Earth (Directly Buried)

Based on Ambient Temperature of 30°C (86°F)

Size	*Temperature Rating of Conductor, See Table 310-13*						*Size*
	60°C (140°F)	*75°C (167°F)*	*90°C (194°F)*	*60°C (140°F)*	*75°C (167°F)*	*90 °C (194°F)*	
AWG kcmil	*Types †TW, †UF*	*Types †FEPW, †RH, †RHW, †THHW, †THW, †THWN, †XHHW, †USE, †ZW*	*Types TBS, SA, SIS, †FEP, †FEPB, MI, †RHH, RHW-2, †THHN, †THHW, †THW-2, †THWN-2 †USE-2, XHH, †XHHW, XHHW-2, ZW-2*	*Types †TW, †UF*	*Types †RH, †RHW, †THHW, †THW, †THWN, †XHHW, †USE*	*Types TA, TBS, SA, SIS, †THHN, †THHW, THW-2, THWN-2, †RHH, RHW-2, USE-2, XHH, XHHW, XHHW-2, ZW-2*	*AWG kcmil*
	Copper			*Aluminum or Copper-Clad Aluminum*			
18	....	....	14	....	....	....	....
16	....	....	18	....	....	....	....
14	20†	20†	25†	....	....	....	....
12	25†	25†	30†	20†	20†	25†	12
10	30	35†	40†	25	30†	35†	10
8	40	50	55	30	40	45	8

6	55	65	75	40	50	60	6
4	70	85	95	55	65	75	4
3	85	100	110	65	75	85	3
2	95	115	130	75	90	100	2
1	110	130	150	85	100	115	1
1/0	125	150	170	100	120	135	1/0
2/0	145	175	195	115	135	150	2/0
3/0	165	200	225	130	155	175	3/0
4/0	195	230	260	150	180	205	4/0
250	215	255	290	170	205	230	250
300	240	285	320	190	230	255	300
350	260	310	350	210	250	280	350
400	280	335	380	225	270	305	400
500	320	380	430	260	310	350	500
600	355	420	475	285	340	385	600
700	385	460	520	310	375	420	700
750	400	475	535	320	385	435	750
800	410	490	555	330	395	450	800
900	435	520	585	355	425	480	900

TABLE 310-16 (continued)

Size	*Temperature Rating of Conductor, See Table 310-13*						*Size*
	60°C (140°F)	*75°C (167°F)*	*90°C (194°F)*	*60°C (140°F)*	*75°C (167°F)*	*90°C (194°F)*	
AWG kcmil	*Types †TW, †UF*	*Types †FEPW, †RH, †RHW, †THHW, †THW, †THWN, †XHHW, †USE, †ZW*	*Types TBS, SA, SIS, †FEP, †FEPB, MI, †RHH, RHW-2, †THHN, †THHW, †THW-2, †THWN-2 †USE-2, XHH, †XHHW, XHHW-2, ZW-2*	*Types †TW, †UF*	*Types †RH, †RHW, †THHW, †THW, †THWN, †XHHW, †USE*	*Types TA, TBS, SA, SIS, †THHN, †THHW, THW-2, THWN-2, †RHH, RHW-2, USE-2, XHH, XHHW, XHHW-2, ZW-2*	*AWG kcmil*
	Copper			*Aluminum or Copper-Clad Aluminum*			
1000	455	545	615	375	445	500	1000
1250	495	590	665	405	485	545	1250
1500	520	625	705	435	520	585	1500
1750	545	650	735	455	545	615	1750
2000	560	665	750	470	560	630	2000

Correction Factors

Ambient Temp. °C	*For ambient temperatures other than 30°C (86°F), multiply the allowable ampacities shown above by the appropriate factor shown below.*						Ambient Temp. °F
21-25	1.08	1.05	1.04	1.08	1.05	1.04	70-77
26-30	1.00	1.00	1.00	1.00	1.00	1.00	78-86
31-35	.91	.94	.96	.91	.94	.96	87-95
36-40	.82	.88	.91	.82	.88	.91	96-104
41-45	.71	.82	.87	.71	.82	.87	105-113
46-50	.58	.75	.82	.58	.75	.82	114-122
51-55	.41	.67	.76	.41	.67	.76	123-131
56-60	....	.58	.71	....	.58	.71	132-140
61-70	....	.33	.58	....	.33	.58	141-158
71-80	....	....	.41	....	....	.41	159-176

† Unless otherwise specifically permitted elsewhere in this Code, the overcurrent protection for conductor types marked with an obelisk (†) shall not excceed 15 amperes for 14 AWG, 20 amperes for 12 AWG, and 30 amperes for 10 AWG copper; or 15 amperes for 12 AWG and 25 amperes for 10 AWG aluminum and copper-clad aluminum after any correction factors for ambient temperature and number of conductors have been applied.

TABLE 310-17 Allowable Ampacities of Single Insulated Conductors, Rated 0–2000 Volts, in Free Air

Based on Ambient Temperature of 30°C (86°F)

Size	*Temperature Rating of Conductor, See Table 310-13*						*Size*
	60°C (140°F)	*75°C (167°F)*	*90°C (194°F)*	*60°C (140°F)*	*75°C (167°F)*	*90 °C (194°F)*	
AWG kcmil	*Types †TW, †UF*	*Types †FEPW, †RH, †RHW, †THHW, †THW, †THWN, †XHHW, †ZW*	*Types TA, TBS, SA, SIS, †FEP, †FEPB, MI, †RHH, RHW-2, †THHN, †THHW, †THW-2, †THWN-2 †USE-2, XHH, †XHHW, XHHW-2, ZW-2*	*Types †TW, †UF*	*Types †RH, †RHW, †THHW, †THW, †THWN, †XHHW*	*Types TA, TBS, SA, SIS, †THHN, †THHW, THW-2, THWN-2, †RHH, RHW-2, USE-2, XHH, †XHHW, XHHW-2, ZW-2*	*AWG kcmil*
	Copper			*Aluminum or Copper-Clad Aluminum*			
18	....	....	18	....	....	....	....
16	....	....	24	....	....	....	....
14	25†	30†	35†	....	....	....	....
12	30†	35†	40†	25†	30†	35†	12
10	40†	50†	55†	35†	40†	40†	10
8	60	70	80	5	55	60	8

6	80	95	105	60	75	80	6
4	105	125	140	80	100	110	4
3	120	145	165	95	115	130	3
2	140	170	190	110	135	150	2
1	165	195	220	130	155	175	1
1/0	195	230	260	150	180	205	1/0
2/0	225	265	300	175	210	235	2/0
3/0	260	310	350	200	240	275	3/0
4/0	300	360	405	235	280	315	4/0
250	340	405	455	265	315	355	250
300	375	445	505	290	350	395	300
350	420	505	570	330	395	445	350
400	455	545	615	355	425	480	400
500	515	620	700	405	485	545	500
600	575	690	780	455	540	615	600
700	630	755	855	500	595	675	700
750	655	785	885	515	620	700	750
800	680	815	920	535	645	725	800
900	730	870	985	580	700	785	900

TABLE 310-17 (continued)

Size	Temperature Rating of Conductor, See Table 310-13						Size
	60°C (140°F)	75°C (167°F)	90°C (194°F)	60°C (140°F)	75°C (167°F)	90°C (194°F)	
AWG kcmil	Types †TW, †UF	Types †FEPW, †RH, †RHW, †THHW, †THW, †THWN, †XHHW, †ZW	Types TA, TBS, SA, SIS, †FEP, †FEPB, MI, †RHH, RHW-2, †THHN, †THHW, †THW-2, †THWN-2 †USE-2, XHH, †XHHW, XHHW-2, ZW-2	Types †TW, †UF	Types †RH, †RHW, †THHW, †THW, †THWN, †XHHW	Types TA, TBS, SA, SIS, †THHN, †THHW, THW-2, THWN-2, †RHH, RHW-2, USE-2, XHH, †XHHW, XHHW-2, ZW-2	AWG kcmil
	Copper			Aluminum or Copper-Clad Aluminum			
1000	780	935	1055	625	750	845	1000
1250	890	1065	1200	710	855	960	1250
1500	980	1175	1325	795	950	1075	1500
1750	1070	1280	1445	875	1050	1185	1750
2000	1155	1385	1560	960	1150	1335	2000

Correction Factors

Ambient Temp. °C	*For ambient temperatures other than 30°C (86°F), multiply the allowable ampacities shown above by the appropriate factor shown below.*						Ambient Temp. °F
21-25	1.08	1.05	1.04	1.08	1.05	1.04	70-77
26-30	1.00	1.00	1.00	1.00	1.00	1.00	79-86
31-35	.91	.94	.96	.91	.94	.96	88-95
36-40	.82	.88	.91	.82	.88	.91	97-104
41-45	.71	.82	.87	.71	.82	.87	106-113
46-50	.58	.75	.82	.58	.75	.82	115-122
51-55	.41	.67	.76	.41	.67	.76	124-131
56-60	....	.58	.71	....	.58	.71	133-140
61-70	....	.33	.58	....	.33	.58	142-158
71-80	....	....	.41	....	....	.41	160-176

† Unless otherwise specifically permitted elsewhere in this Code, the overcurrent protection for conductor types marked with an obelisk (†) shall not excceed 15 amperes for 14 AWG, 20 amperes for 12 AWG, and 30 amperes for 10 AWG copper; or 15 amperes for 12 AWG and 25 amperes for 10 AWG aluminum and copper-clad aluminum.

TABLE 310-18 Allowable Ampacities of Three Single Insulated Conductors Rated 0–2000 Volts, 150 to 250° C (302 to 482° F), in Raceway or Cable

Based on Ambient Air Temperature of 40°C (104°F)

Size	*Temperature Rating of Conductor, See Table 310-13*				*Size*
	150°C (302°F)	*200°C (392°F)*	*250°C (482°F)*	*150°C (302°F)*	
AWG kcmil	*Type Z*	*Types FEP, FEPB, PFA*	*Types PFAH, TFE*	*Type Z*	*AWG kcmil*
	Copper		*Nickel or Nickel-Coated Copper*	*Aluminum or Copper-Clad Aluminum*	
14	34	36	39	....	14
12	43	45	54	30	12
10	55	60	73	44	10
8	76	83	93	57	8

6	96	110	117	75	6
4	120	125	148	94	4
3	143	152	166	109	3
2	160	171	191	124	2
1	186	197	215	145	1
1/0	215	229	244	169	1/0
2/0	251	260	273	198	2/0
3/0	288	297	308	227	3/0
4/0	332	346	361	260	4/0
250	....	....	....	....	205
300	....	....	....	....	300
350	....	....	....	....	350
400	....	....	....	....	400
500	....	....	....	....	500
600	....	....	....	....	600
700	....	....	....	....	700
750	....	....	....	....	750
800	....	....	....	....	800
1000	....	....	....	....	1000
1500	....	....	....	....	1500
2000	....	....	....	....	2000

TABLE 310-18 (continued)

Correction Factors					
Ambient Temp. °C	*For ambient temperatures other than 40°C (104°F), multiply the allowable ampacities shown above by the appropriate factor shown below*				*Ambient Temp. °F*
41-50	.95	.97	.98	.95	105-122
51-60	.90	.94	.95	.90	123-140
61-70	.85	.90	.93	.85	141-158
71-80	.80	.87	.90	.80	159-176
81-90	.74	.83	.87	.74	177-194
91-100	.67	.79	.85	.67	195-212
101-120	.52	.71	.79	.52	213-248
121-140	.30	.61	.72	.30	240-284
141-160	....	.50	.65	....	285-320
161-180	....	.35	.58	....	321-356
181-200	....	....	.49	....	357-392
201-225	....	....	.35	....	393-437

conductors of 0 through 2000 V are shown in Tables 310-16 through 310-19 and accompanying notes. The ratings for solid dielectric insulated conductors of 2001 through 35,000 V are shown in Tables 310-67 through 310-84. Tables 310-16 through 310-19 are application tables for use in accordance with Article 220. It is suggested that the reader become familiar with this section in its entirety in the Code. For convenience, only Tables 310-16 through 310-19 are shown here.

If more than one ampacity is determined from the tables or calculations, the lowest value must be used. There is one exception for adjacent portions of a circuit under certain circumstances.

Electrical ducts noted in Article 310 are any conduits noted in Chapter 3 for use underground; round raceways listed for use underground; embedded in concrete or earth.

With engineering supervision the following formula can be used to calculate ampacities:

$$I = \sqrt{\frac{TC - (TA + DELTA\ TD)}{RDC(1 + YC)RCA}}$$

where

TC = conductor temperature in degrees C
TA = ambient temperature in degrees C
DELTA TD = dielectric loss temperature rise
RDC = dc resistance of conductor at temperature TC
YC = component ac resistance resulting from skin effect and proximity effect
RCA = effective thermal resistance between conductor and surrounding ambient

Refer to Appendix B for examples of formula application. The ampacities provided by this section are based on temperature alone and do not take voltage drop into consideration. Conductors of circuits as defined in Article 100, sized to prevent a voltage drop exceeding 5%, will provide reasonable efficiency of operation.

Notes to Ampacity Tables of 0-2000 Volts

1. Explanation of Tables

For explanation of Type Letters, and for recognized size of conductors for the various conductor insulations, see Section

TABLE 310-19 Allowable Ampacities of Single Insulated Conductors Rated 0 through 2000 Volts, 150° to 250°C (302° to 482°F), in Free Air

Based on Ambient Air Temperature of 40°C (104°F)

Size	*Temperature Rating of Conductor, See Table 310-13*					*Size*
	150°C (302°F)	*200°C (392°F)*		*250°C (482°F)*	*150°C (302°F)*	
AWG kcmil	*Type Z*	*Types FEP, FEPB, PFA*	*Bare or Covered Conductors*	*Types PFAH, TFE*	*Type Z*	*AWG kcmil*
	Copper			*Nickel or Nickel-Coated Copper*	*Aluminum or Copper-Clad Aluminum*	
14	46	54	30	59	....	14
12	60	68	35	78	47	12
10	80	90	50	107	63	10
8	106	124	70	142	83	8

6	155	165	95	205	112	6
4	190	220	125	278	148	4
3	214	252	150	327	170	3
2	255	293	175	381	198	2
1	293	344	200	440	228	1
1/0	339	399	235	532	263	1/0
2/0	390	467	275	591	305	2/0
3/0	451	546	320	708	351	3/0
4/0	529	629	370	830	411	4/0
250	....	....	415	....	....	250
300	....	....	460	....	....	300
350	....	....	520	....	....	350
400	....	....	560	....	....	400
500	....	....	635	....	....	500
600	....	....	710	....	....	600
700	....	....	780	....	....	700
750	....	....	805	....	....	750
800	....	....	835	....	....	800
900	....	....	865	....	....	900
1000	....	....	895	....	....	1000
1500	....	....	1205	....	....	1500
2000	....	....	1420	....	....	2000

TABLE 310-19 (continued)

	Correction Factors					
Ambient Temp. °C	*For ambient temperatures other than 40°C (104°F), multiply the allowable ampacities shown above by the appropriate factor shown below*					*Ambient Temp. °F*
41-50	.95	.97	....	.98	.95	105-122
51-60	.90	.94	....	.95	.90	123-140
61-70	.85	.90	....	.93	.85	141-158
71-80	.80	.87	....	.90	.80	159-176
81-90	.74	.83	....	.87	.74	177-194
91-100	.67	.79	....	.85	.67	195-212
101-120	.52	.71	....	.79	.52	213-248
121-140	.30	.61	....	.72	.30	249-284
141-160	....	.50	....	.65	....	285-320
161-180	....	.35	....	.58	....	321-356
181-200	....	....	....	.49	....	357-392
201-225	....	....	....	.35	....	393-437

310-13. For installation requirements, see Sections 310-1 through 310-10, and the various articles of this Code. For flexible cords, see Tables 400-4, 400-5(A), and 400-5(B).

3. 120/240 Volts, Three-Wire, Single-Phase Dwelling Services and Feeders

For dwelling units, conductors, as listed below, shall be permitted to be utilized as 120/240-V, 3-wire, single-phase service-entrance conductors, service lateral conductors, and feeder conductors that serve as the main power feeder to a dwelling unit and are installed in raceway or cable with or without an equipment grounding conductor. For applications of this note, the feeder conductors shall not be required to be larger than its service-entrance conductors. The grounded conductor shall be permitted to be smaller than the ungrounded conductors, provided that the requirements of Sections 215-2, 220-22, and 230-42 are met.

5. Bare Conductors or Covered Conductors

Where bare or covered conductors are used with insulated conductors, their allowable ampacities shall be limited to that permitted for the adjacent insulated conductors.

Conductor Types and Sizes
RH-RHH-RHW-THHW-THW-THWN-THHN-XHHW-USE

Copper	*Aluminum and Copper-Clad AL*	*Service or Feeder Ratings in Amps*
AWG	AWG	
4	2	100
3	1	110
2	1/0	125
1	2/0	150
1/0	3/0	175
2/0	4/0	200
3/0	250 kcmil	225
4/0	300 kcmil	250
250 kcmil	350 kcmil	300
350 kcmil	500 kcmil	350
400 kcmil	600 kcmil	400

6. Mineral-Insulated, Metal-Sheathed Cable

The temperature limitation on which the ampacities of mineral-insulated, metal-sheathed cable are based is determined by the insulating materials used in the end seal. Termination fittings incorporating unimpregnated, organic, insulating materials are limited to 90°C (194°F) operation.

7. Type MTW Machine Tool Wire

(FPN): For the ampacities of Type MTW wire, see Table 13-5(a) in the Electrical Standard for Industrial Machinery, NFPA 79-1994.

8. Adjustment Factors

(a) More than Three Current-Carrying Conductors in a Raceway or Cable. Where the number of current-carrying conductors in a raceway or cable exceeds three, the ampacities shall be reduced as shown in the following table:

Number of Conductors	*Percent of Values in Tables as Adjusted for Ambient Temperature if Necessary*
4 through 6	80
7 through 9	70
10 through 20	50
21 through 30	45
31 through 40	40
41 through 60	35

Where single conductors or multiconductor cables are stacked or bundled longer than 24 in. (610 mm) without maintaining spacing and are not installed in raceways, the allowable ampacity of each conductor shall be reduced as shown in the above table.

Exception No. 1. When conductors of different systems, as provided in Section 300-3, are installed in a common raceway or cable the derating factors shown above shall apply to the number of power and lighting (Articles 210, 215, 220, and 230) conductors only.

Exception No. 2. For conductors installed in cable trays, the provisions of Section 318-11 shall apply.

Exception No. 3. Derating factors shall not apply to conductors in nipples having a length not exceeding 24 in. (610 mm).

Exception No. 4. Derating factors shall not apply to underground conductors entering or leaving an outdoor trench if those conductors have physical protection in the form of rigid metal conduit, intermediate metal conduit or rigid nonmetallic conduit having a length not exceeding 10 ft. (3.05 m) and the number of conductors does not exceed 4.

Exception No. 5. For other loading conditions, adjustment factors and ampacities shall be permitted to be calculated under Section 310-15(b).

(FPN): See Appendix B, Table B-310-11 for adjustment factors for more than three current-carrying conductors in a raceway or cable with load diversity.

(b) More than One Conduit, Tube, or Raceway. Spacing between conduits, tubing, or raceways shall be maintained.

9. Overcurrent Protection

Where the standard ratings and settings of overcurrent devices do not correspond with the rating and settings allowed for conductors, the next higher standard rating and setting shall be permitted.

Exception. As limited in Section 240-3.

10. Neutral Conductor

(a) A neutral conductor that carries only the unbalanced current from other conductors of the same circuit need not be counted when applying the provisions of Note 8.

(b) In a three-wire circuit consisting of two-phase wires and the neutral of a four-wire, three-phase wye-connected system, a common conductor carries approximately the same current as the line to neutral load currents of the other conductors and shall be counted when applying the provisions of Note 8.

(c) On a four-wire, three-phase wye circuit where the major portion of the load consists of nonlinear loads there are harmonic currents present in the neutral conductor and the neutral shall be considered to be a current-carrying conductor.

11. Grounding or Bonding Conductor

A grounding or bonding conductor shall not be counted when applying the provisions of Note 8.

ARTICLE 318
CABLE TRAYS

318-1 Scope

318-2 Definition

318-3 Uses Permitted

(a) **Wiring Methods.** The following types of cables can be used in cable trays provided they meet specific requirements found in the descriptions for each cable. The reader should refer to specific articles in the *NEC*® for additional requirements. The cables are:

1. Armored cable
2. Mineral-insulated, metal-sheathed cable
3. Power-limited tray cable
4. Metal-clad cable
5. Nonmetallic-sheathed cable
6. Multiconductor underground feeder and branch circuit conductor cable
7. Multiconductor service-entrance cable
8. Power and control tray cable
9. Other factory-assembled, multiconductor control, signal, or power cables approved for this purpose
10. Electrical nonmetallic tubing
11. Intermediate metal conduit
12. Rigid nonmetallic conduit
13. Rigid metal conduit
14. Electrical metallic tubing
15. Flexible metal conduit
16. Flexible metallic tubing
17. Liquidtight flexible metal conduit and liquidtight flexible nonmetallic conduit
18. Optical fiber cables
19. Instrumentation tray cable

(b) In industrial establishments. Wiring methods of 318-3(a) can be used in accordance with their respective article's requirements. Where only qualified maintenance personnel are available, certain cables are permitted in ladder, ventilated trough, and ventilated channel-type cable trays. Single-conductor cable must be 1/0 or larger and be permitted to be used in cable tray. If the cable is sized 1/0 through 4/0, it must be installed in ladder-type cable tray with a maximum rung spacing of 9 in. or in ventilated trough cable tray. Welding cable can be installed in accordance with Article 630, Part E. Equipment grounding single conductors can be insulated, covered, or bare No. 4 or larger. Multiconductor cable type MV can also be installed as noted above. In all cases, the cable must be sunlight resistant if exposed to the sun.

(c) Equipment Grounding Conductors

(d) Hazardous (Classified) Locations

(e) Nonmetallic Cable Tray

318-4 Uses Not Permitted

Cable trays cannot be used where they are exposed to severe physical injury, in hoistways, and in environmental air-handling spaces. An exception is made in the latter case where used to support a wiring method noted in Section 300-22(c).

318-5 Construction Specifications

Cable trays must have proper strength and rigidity, smooth edges, be protected against corrosion, have side rails, and contain the proper fittings. Flame-retardant material must be used for nonmetallic cable tray.

318-6 Installation

The cable tray system must be installed completely as a system before the cables are installed, have the proper supports, and have covers where required. Multiconductor cables can be installed in the same tray if 600 V or less. Cables over 600 V cannot be installed with cables under 600 V, with two exceptions. The cable trays must be exposed and accessible and have room around them for adequate access. They can go through partitions, walls, and floors if installed in accordance with Section 300-21 (proper and approved firestopping methods used). Incidental support from cable tray is permitted where approved and designed to support the load.

318-7 Grounding

Proper grounding must be attained in compliance with Article 250. The entire system must be grounded. Minimum size steel and aluminum trays are indicated in Table 318-7(b)(2) of the *NEC*®.

318-8 Cable Installation

(a) Splices are permitted if approved methods are used, they are accessible, and are not over the side rails.

(b) Cables must be fastened in sloping or vertical runs.

(c) If the cable is installed in a bushed conduit and tubing for protection or support, a box is not required.

(d) If installed in parallel, the conductors must be grouped with one conductor per phase in each group. Single conductors must be bound to circuit groups.

318-9 Number of Multiconductor Cables, Rated 2000 Volts, Nominal, or Less, in Cable Trays

Refer to this section in the Code and Table 318-9.

318-10 Number of Single Conductor Cables, Rated 2000 Volts or Less, in Cable Trays

Refer to this section in the Code and Table 318-10.

318-11 Ampacity of Cables Rated 2000 Volts or Less in Cable Trays

(a) Multiconductor Cables. Multiconductor cables, rated 2000 V or less, installed as per Section 318-9 are to follow ampacities in Tables 310-16 and 310-18. The derating factors of Article 310, Note 8(a) of the Notes to Ampacity Tables of 0 to 2000 volts only apply to multiconductor cables of three or more current carrying conductors in cable, not in cable tray. If the tray is covered for more than 6 ft. with a solid unventilated cover, the ampacities allowed are 95% of those shown in the tables. An additional exception is allowed for a single layer in uncovered trays.

(b) Single Conductor Cables. The derating factors of Article 310, Note 8(a) of the Notes to Ampacity Tables of 0 to 2000 volts don't apply to cables in cable tray. Single conductor or single conductors cabled together, rated 2000 V or less, properly installed in accordance with the Code, are to use ampacities as follows:

1. Cables 600 kcmil and larger in uncovered trays use not more than 75% of the ampacities in Tables 310-17 and 310-19. If the tray is covered for more than 6 ft. with unventi-

lated covers, use a maximum 70% of the ampacities in the tables.

2. Cables 1/0 through 500 kcmil in uncovered trays use not more than 65% of the ampacities in Tables 310-17 and 310-19. If the tray is covered for more than 6 ft. with unventilated covers, use a maximum 60% of the ampacities in the tables.

3. If single conductor cables are placed in a single layer with at least one cable diameter separation, the tray is uncovered and the cable is 1/0 or larger, the ampacities in Tables 310-17 and 310-19 may be used.

4. Special allowance when installed in uncovered tray with a triangular configuration.

318-12 Number of Type MV and MC Cables (2001 Volts or Over) in Cable Trays

318-13 Ampacity of Type MV and Type MC Cables (2001 Volts or Over) in Cable Trays

ARTICLE 320
OPEN WIRING ON INSULATORS

320-1 Definition

This type of wiring is exposed with a single insulated conductor using an insulator (knob, cleat, tube, or flexible tubing) for protection and support.

320-2 Other Articles

Must also comply with Articles 225 and 320 as well as other Code articles that apply.

320-3 Uses Permitted

This type of wiring can be used only where the nominal voltage does not exceed 600 V, in industrial or agricultural structures, both indoors or outdoors, both wet or dry locations, for services, and where corrosive vapors are present.

320-5 Conductors

320-6 Conductor Supports

320-7 Mounting of Conductor Supports

320-8 Tie Wires

320-10 Flexible Nonmetallic Tubing

320-11 Through Walls, Floors, Wood Cross Members, etc.

320-12 Clearance from Piping, Exposed Conductors, etc.

320-13 Entering Spaces Subject to Dampness, Wetness, or Corrosive Vapors

320-14 Protection from Physical Damage

320-15 Unfinished Attics and Roof Spaces

320-16 Switches

ARTICLE 321
MESSENGER SUPPORTED WIRING

321-1 Definition

This type of wiring is exposed using a messenger wire as a support for insulated conductors. These conductors are usually in the form of a type of cable. In some instances in industrial establishments, insulated wire can be used as well as cable.

321-2 Other Articles

Must also comply with Articles 225 and 300 as well as other Code articles that apply.

321-3 Uses Permitted

321-4 Uses Not Permitted

321-5 Ampacity

321-6 Messenger Support

321-7 Grounding

321-8 Conductor Splices and Taps

ARTICLE 324
CONCEALED KNOB-AND-TUBE WIRING

324-1 Definition

This type of wiring is a single insulated conductor using knobs, tubes, and flexible nonmetallic tubing for support and protection.

324-2 Other Articles

It must also comply with Article 300 as well as other Code articles that apply.

324-3 Uses Permitted

It can only be used to extend an existing installation or elsewhere with special permission from the authorities having jurisdiction and in hollow spaces of walls and ceilings and unfinished attic and roof spaces (refer to Section 324-11 in the *NEC®*).

324-4 Uses Not Permitted

It is not allowed in commercial garages, motion picture studios, classified hazardous locations, and theaters and similar locations, or in hollow spaces of walls, ceilings, and attics under certain conditions.

324-5 Conductors

324-6 Conductor Supports

324-7 Tie Wires

324-8 Conductor Clearances

324-9 Through Walls, Floors, Wood Cross Members, etc.

324-10 Clearance from Piping, Exposed Conductors, etc.

324-11 Unfinished Attics and Roof Spaces

324-12 Splices

324-13 Boxes

324-14 Switches

ARTICLE 325
INTEGRATED GAS SPACER CABLE
Type IGS

A. General

325-1 Definition

325-2 Other Articles

325-3 Uses Permitted

325-4 Uses Not Permitted

B. Installation

325-11 Bending Radius

325-12 Bends

325-13 Fittings

325-14 Ampacity

C. Construction Specifications

325-20 Conductors

325-21 Insulation

325-22 Conduit

325-23 Grounding

325-24 Marking

ARTICLE 326
MEDIUM-VOLTAGE CABLE
Type MV

326-1 Definition

This cable is rated above 2000 V and is a single or multiconductor solid dielectric insulated cable.

326-2 Other Articles

It must also comply with Articles 300, 305, 310, 318, 501, and 710 as well as other Code articles that apply.

326-3 Uses Permitted

It can be used in dry or wet locations, in raceways, directly buried, and in messenger supported cable in systems up to 35,000 V. Refer to Sections 318-3(b)(1) and 710-4(b).

326-4 Uses Not Permitted

It cannot be used in cable trays or direct sunlight unless approved for those purposes.

326-5 Construction

326-6 Ampacity

326-7 Markings

ARTICLE 328
FLAT CONDUCTOR CABLE
Type FCC

328-1 Scope

The wiring covered in this article is a field-installed system to be used for branch circuits under carpet squares.

328-2 Definitions

The reader is referred to the Code for these definitions.

328-3 Other Articles

It must also comply with Articles 210, 220, 240, 250, and 300 as well as other Code articles that apply.

328-4 Uses Permitted

It can be used for general-purpose, appliance, and individual branch circuits; on floors that are hard, sound, or smooth made of materials like or similar to concrete, ceramic, composition, and wood; on walls in surface metal raceways; in damp locations; and on heated floors where the temperature is higher than 86°F if approved for those temperatures.

328-5 Uses Not Permitted

It cannot be used outdoors, in wet locations, in classified hazardous locations, where there are corrosive vapors present, or in residential, school, or hospital buildings.

328-6 Branch-Circuit Ratings

The voltage ratings are up to 150 V to ground and up to 300 V between ungrounded conductors. The current ratings are up to 20 A for general-purpose and appliance branch circuits and up to 30 A for individual branch circuits.

B. Installation

328-10 Coverings

When on a floor, it must be covered with carpet squares. Release-type adhesive only is permitted.

328-11 Cable Connections and Insulating Ends

328-12 Shields

328-13 Enclosure and Shield Connections

328-14 Receptacles

They must be approved for this system.

328-15 Connection to Other Systems

Use approved transition assemblies.

328-16 Anchoring

328-17 Crossings

328-18 System Height

If it exceeds 0.09 in. it must be feathered or tapered at the edges to floor level.

328-19 FCC Systems Alterations

328-20 Polarization of Connections

C. Construction

328-30 Type FCC Cable

Three, four, or five flat copper conductors; one is an equipment grounding conductor.

328-31 Markings

328-32 Conductor Identification

328-33 Corrosion Resistance

328-34 Insulation

328-35 Shields

328-36 Receptacles and Housings

328-37 Transition Assemblies

ARTICLE 330
MINERAL-INSULATED, METAL-SHEATHED CABLE
Type MI

A. General

330-1 Definition

Type MI cable is factory assembled with one or more conductors using a highly compressed refractory mineral insulation. The entire assembly is covered with a gastight and liquidtight continuous copper or alloy steel sheath.

330-2 Other Articles

It must also comply with Article 300 as well as other Code articles that apply.

330-3 Uses Permitted

It can be used for services, feeders, and branch circuits; power, control, lighting, and signal circuits; wet, dry, or continuously moist locations; exposed or concealed; indoors and outdoors; embedded in concrete, plaster, fill, and other masonry; hazardous locations; underground when protected; corrosive locations that will not harm the sheath; and where exposed to gasoline and oil.

330-4 Uses Not Permitted

Where exposed to corrosive materials that will harm the sheath unless it is properly protected.

B. Installation

330-10 Wet Locations

330-11 Through Joists, Studs or Rafters

330-12 Supports

330-13 Bends

330-14 Fittings

330-15 Terminal Seals

330-16 Single Conductors

C. Construction Specifications

330-20 Conductors

Solid copper or nickel-clad copper.

330-21 Insulation

Highly compressed refractory mineral.

330-22 Outer Sheath

Continuous construction that ensures mechanical protection, ground path, and a moisture seal.

ARTICLE 331
ELECTRICAL NONMETALLIC TUBING

A. General

331-1 Definition

331-2 Other Articles

331-3 Uses Permitted

331-4 Uses Not Permitted

B. Installation

331-5 Size

331-6 Number of Conductors in Tubing

331-7 Trimming

331-8 Joints

331-9 Bends—How Made

331-10 Bends—Number in One Run

331-11 Supports

331-12 Boxes and Fittings

331-13 Splices and Taps

331-14 Bushings

C. Construction Specifications

331-15 General

ARTICLE 333
ARMORED CABLE
Type AC
A. General

333-1 Definition

This type of cable is composed of insulated conductors in a flexible metal enclosure. Refer to Section 333-19.

333-2 Other Articles

It must also comply with Article 300 as well as other Code articles that apply.

333-3 Uses Permitted

It can be used for feeders and branch circuits, exposed or concealed and in cable trays where so identified. It can not be subject to physical damage. It is permitted in dry locations and in plaster finish on brick or other masonry except in damp or wet locations. It can also be used in the voids of masonry block or tile walls when the walls are not exposed, below grade or subject to excessive moisture or dampness.

333-4 Uses Not Permitted

It cannot be used where prohibited in the Code. This includes in theaters or similar locations except as noted in Article 518; motion picture studios; in hazardous locations except where noted in Sections 501-4(b), 502-4(b), and 504-20; where exposed to corrosive fumes; on cranes or hoists except as noted in Section 610-11 Exceptions No. 3; in storage battery rooms; in hoistways or on elevators except as noted in Section 620-11; or in commercial garages where noted in Article 511.

B. Installation

333-7 Support

333-8 Bending Radius

They must be made so that the cable will not be damaged and the radius of the curve of the inner edge can not be less than five times the diameter of the cable.

333-9 Boxes and Fittings

333-10 Through or Parallel to Framing Member

333-11 Exposed work

333-12 In Accessible Attics

C. Construction Specifications

333-19 Construction

The cable consists of insulated conductors (in accordance with Section 333-20) enclosed in a flexible metal tape armor with a copper or aluminum bonding strip inside and in contact with the armor.

333-20 Conductors

333-21 Grounding

333-22 Marking

ARTICLE 334
METAL-CLAD CABLE

A. General

334-1 Definition

Metal-clad cable is Type MC. It is a cable in which the conductors are enclosed in a metal sheath, insulated, and the entire cable is put together at the factory.

334-2 Other Articles

This type of cable must be in accordance with Article 300 as well as other Code articles that apply.

334-3 Uses Permitted

This cable can be used for almost all types of applications, including wet locations, provided that it will not sustain physical damage and is not prohibited in another part of the *NEC®*. A listing is given in the *NEC®*. In wet locations one of the following must be adhered to: moisture will not get through the metal cover, a lead sheath that will not let moisture through is put under the metal cover, or the insulated conductors enclosed by the metal cover are listed for a wet location. Refer to this section in the Code.

334-4 Uses Not Permitted

This cable cannot be used in a location where it is subject to corrosive conditions that are destructive. The *NEC®* gives as examples direct buried cable in earth or concrete, under fills, chlorides, caustic alkalis, and the like. If the metal cover will withstand these reactions, it can be used in such locations.

B. Installation

334-10 Installation

Type MC cable must be installed in accordance with the requirements of Articles 300, 710, 725, and Section 770-52. When installed in cable tray, direct buried, as service-entrance cable outside buildings or as aerial cable, in accessible attics, and through or parallel to joists, studs, and rafters, you must refer to the appropriate articles in the *NEC®*. It must be held at least every 6 ft. and within 12 in. of every outlet box, device box, junction box, cabinet or terminal fitting. There are three exceptions.

334-11 Bending Radius

The following are the allowable radii of the curve of the inner edge: for shielded conductors, use the greater of either 12 times the diameter of one conductor or 7 times the overall diameter of the cable; for interlocked armor or corrugated sheath, 7 times the diameter of the sheath; for smooth sheath, use the following:

Cable up to ¾ in. diameter: 10 times diameter
Cable between ¾ and 1½ in. diameter: 12 times diameter
Cable larger than 1½ in. diameter: 15 times diameter

334-12 Fittings

334-13 Ampacity

You must use the ampacities in accordance with Section 310-15. There are two exceptions.

C. Construction Specifications

334-20 Conductors

Copper, aluminum, or copper-clad aluminum must be used, either stranded or solid. The smallest sizes allowed are No. 18 for copper and No. 12 for aluminum or copper-clad aluminum.

334-21 Insulation

334-22 Metallic Sheath

The sheath must be either smooth, welded, and corrugated or interlocking.

334-23 Grounding

334-24 Marking

ARTICLE 336
NONMETALLIC-SHEATHED CABLE
Types NM, NMC, and NMS

A. General

Note: This Article has been completely rewritten. The reader is cautioned to refer to the *NEC*® for the complete information.

336-1 Scope

336-2 Definition

Nonmetallic-sheathed cable is either Type NM or NMC. It is a cable in which the insulated cables are enclosed in a nonmetallic material that is moisture-resistance and flame-retardant. The cable is put together at the factory. Type NMS is a factory assembly of power and remote control signaling conductors in a common jacket with the same properties as NM and NMC.

336-3 Other Articles

Note other applicable articles, such as Article 300 and Article 316.

336-4 Uses Permitted

This type of cable can be used only in buildings that do not exceed three floors above grade. Type NM and NMS cable can be used either exposed or concealed in dry locations. It can be used in the voids of masonry or tile if not a damp location. Type NMC cable can be used where type NM cable is used and also in moist damp and corrosive areas and in walls of masonry block and the like. Type NMS can be used where permitted by Article 280. Refer to the *NEC*® for a complete list.

336-5 Uses Not Permitted

Type NM, NMC, and NMS cannot be used as service-entrance cable, in any structure exceeding 3 stories above grade, in commercial garages, in motion picture studios, in theaters, and the like unless otherwise provided in the *NEC*®, in hoistways, storage battery rooms, hazardous locations, and in poured concrete cement or aggregate. The reader is referred to this section of the *NEC*® for a complete list of locations where types NM, NMC, and NMS are not permitted. Refer to the *NEC*® for a complete list.

B. Installation

336-6 Exposed Work—General

336-9 Through or Parallel to Framing Members

336-16 Bends

336-18 Supports

The cable must be secured at least every 4½ ft. and not more than 12 in. from a fitting, box, or cabinet. Two exceptions noted. Only certain supports are approved.

336-20 Boxes of Insulating Material

336-21 Devices of Insulating Material

336-25 Devices with Integral Enclosures

C. Construction Specifications

336-30 General

336-31 Marking

ARTICLE 338
SERVICE-ENTRANCE CABLE
Types SE and USE

338-1 Definition

This cable is either Type SE or USE and is used primarily for services. Type SE has a flame-retardant and moisture-resistant cover. Type USE, which is the one used underground, has a moisture-resistant cover.

338-2 Uses Permitted as Service-Entrance Conductors

When used as service-entrance conductors, you must follow Article 230. If protected as noted in Section 300-5 Type USE can come above ground outside in meter enclosures or terminations when used as a service lateral.

338-3 Uses Permitted as Branch Circuits or Feeders

(a) Grounded Conductor Insulated. Type SE can be used if all conductors are rubber-covered or thermoplastic type.

(b) Grounded Conductor Not Insulated. Generally, Type SE Cable with a noninsulated grounded conductor cannot be used for branch circuit or feeder wiring in a building. There are exceptions with specific rules for ranges, wall-mounted ovens, counter-mounted cooking units, or clothes dryers.

(c) Temperature Limitations. Do not exceed temperature specified.

338-4 Interior Installation Methods

Review applicable sections of the *NEC®*.

338-5 Marking

338-6 Bends

ARTICLE 339
UNDERGROUND FEEDER AND BRANCH-CIRCUIT CABLE
Type UF

339-1 Description and Marking

This cable is Type UF. Sizes allowed are minimum No. 14 copper or No. 12 aluminum or copper-clad aluminum and maximum No. 4/0. The conductors must be of the moisture-resistant type found in Table 310-13. It must be suitable for branch-circuit wiring and the ampacity to use is listed under 60°C (140°F) in Table 310-16. The overall covering must be flame-retardant, moisture-, fungus-, and corrosion-resistant, and able to be directly buried. The cable can have an insulated or bare conductor only for equipment grounding purposes.

339-2 Other Articles

Conform to other Code articles where applicable, in particular Article 300 and Section 310-13.

339-3 Use

(a) Uses Permitted. In underground, including direct burial, as a feeder or branch circuit as long as the overcurrent requirements of Section 339-4 are adhered to. It can be used for interior wiring in wet, dry, or corrosive areas provided that all other conditions are met. In certain of these cases, as described in Section 690-31, it can not be used for solar photovoltaic systems. Refer to Section 300-5 for underground requirements. When single conductor cables are used, all cables of the feeder or branch circuit have to be run in the same trench or wireway. There are other exceptions.

(b) Uses Not Permitted. This type of cable cannot be used for service entrances, in commercial garages, in theaters, in storage battery rooms, in motion picture studios, in hazardous locations, in hoistways, in poured cement, or where exposed to direct rays of sun unless approved for sunlight installation.

339-4 Overcurrent Protection

339-5 Rated Ampacity

ARTICLE 340
POWER AND CONTROL TRAY CABLE
Type TC

340-1 Definition

This cable is Type TC. It is nonmetallic-sheathed cable used in cable trays, raceways, or supported by messenger wire with insulated conductors. The grounding conductor can be either bare or covered.

340-2 Other Articles

Conform to other Code articles where applicable, in particular Articles 300 and 318.

340-3 Construction

Minimum-size conductors are No. 18 for copper and No. 12 for aluminum and copper-clad aluminum. Maximum size is 1000 kcmil. Conductors of No. 18 and No. 16 sizes are to be in accordance with Section 725-16. No. 14 copper and larger and No. 12 or larger aluminum or copper-clad aluminum are to be of the types shown in Table 310-13 or 310-62. The sheath is to be flame-retardant and nonmetallic. A metallic sheath cannot be put over or under the nonmetallic sheath. If it is installed in a wet area, it must also be moisture- and corrosion-resistant as applicable. There is an exception for fire signaling systems and one when used in accordance with Article 725.

340-4 Uses Permitted

This cable can be used for power, lighting, control, signal and communications circuits; in cable trays, raceways, or supported by messenger wire; in hazardous locations when permitted in Articles 318, 501, 502, and 504; under certain conditions where permitted by Article 725 for Class I circuits; and for nonpower-limited fire alarm circuits in accordance with Section 760-16.

340-5 Uses Not Permitted

Type TC cable cannot be used where exposed to physical damage; on open brackets; where exposed to direct sunlight unless sunlight resistant; and direct buried unless approved for that purpose. There is an exception.

340-6 Marking

340-7 Ampacity

Use Section 402-5 and Section 318-11.

340-8 Bends

ARTICLE 342
NONMETALLIC EXTENSIONS

342-1 Definition

This is a conductor arrangement with two insulated conductors in a nonmetallic jacket or thermoplastic cover.

342-2 Other Articles

342-3 Uses Permitted

Only from an existing 15- or 20-A outlet on a branch circuit. It must be run exposed and dry. For surface extensions the buildings must only be occupied for residential or office use and with height restrictions of Section 336-5(a)(1).

342-4 Uses Not Permitted

This type of extension cannot be used as an aerial cable other than where permitted; in unfinished areas, where there are corrosive vapors, through a floor or partition; and where the voltage between conductors is greater than 150 V for surface work and 300 V for aerial cable.

342-5 Splices and Taps

342-6 Fittings

342-7 Installation

342-8 Marking

ARTICLE 343
NONMETALLIC UNDERGROUND CONDUIT WITH CONDUCTORS

A. General

343-1 Description

This is a factory assembly of conductors or cables in a nonmetallic circular conduit. The conduit has to be moisture and corrosion resistant. It must be capable of being shipped on reels without damage and should be able to withstand abuse.

343-2 Other Articles

Must be in accordance with Article 300.

343-3 Uses Permitted

It can be used encased or embedded in concrete, in cinder fill, for direct burial cable with cover requirements as noted in Tables 300-5 and 710-4(b) under rigid nonmetallic conduit, and in underground locations subject to corrosive action as noted in Section 300-6 and where subject to chemicals when approved for it.

343-4 Uses Not Permitted

It cannot be used in exposed locations, inside buildings with one exception, and in hazardous (classified) locations except as noted in Sections 503-3(a), 504-20, 514-8, 515-5, and 501-4(b).

B. Installation

343-5 Size

It cannot be smaller than ½ in. or larger than 4 in.

343-6 Trimming

343-7 Joints

343-8 Conductor Terminations

343-9 Bushings

343-10 Bends—How Made

They must be made manually, the conduit not be damaged and the internal diameter not be reduced.

343-11 Bends—Number in One Run

The maximum is not more than 360 between termination points.

343-12 Splices and Taps

C. Construction

343-13 General

343-14 Conductors and Cables

343-15 Conductor Fill

Not to exceed number shown in Table 1 of Chapter 9.

343-16 Marking

Must be marked every 10 ft. It also must be identified on a tag at each end of the assembly or on the side of the reel.

ARTICLE 345
INTERMEDIATE METAL CONDUIT

A. General

345-1 Definition

This conduit is a circular metal raceway with all its couplings, etc.

345-2 Other Articles

Must be in accordance with Article 300.

345-3 Uses Permitted

Intermediate metal conduit can be used in most places and under most conditions. With the exception of aluminum fittings and enclosures and steel conduit, dissimilar metals in contact must not be used where possible. It can be used in the earth or in corrosive locations if it has corrosion protection. It can be used in cinder fill only if 2 in. of noncinder concrete is in direct contact, the conduit is 18 in. or more under the fill, or there is proper corrosion protection.

B. Installation

345-5 Wet Locations

Everything must be corrosion resistant (bolts, straps, etc.).

345-6 Size

The conduit used cannot be smaller than ½ in. or larger than 4 in.

345-7 Number of Conductors in Conduit

Refer to Chapter 9, Tables 1 and 4.

345-8 Reaming and Threading

All rough edges must be removed when the conduit is cut. When threaded in the field, a tapered cutting die (¾ in. per foot) must be used.

345-9 Couplings and Connectors

Threadless couplings and connectors can be used. Running threads cannot be used. Use concrete type in concrete and raintight in wet areas.

345-10 Bends—How Made

The conduit must not be injured. Refer to Table 346-10 for the radius of the inner edge for field bends. If the conduit does not have a lead sheath and the bend is made with a bending machine that uses one operation (one shot) the Table 346-10 Exception can be used.

345-11 Bends—Number in One Run

The maximum that is permitted is not more than 360° in bends between pull points.

345-12 Supports

The conduit must be supported at intervals no longer than 10 ft and fastened not more than 3 ft from a box, fitting, etc. This may be extended to 5 ft where structural members prevent fastening within 3 ft. There are four exceptions—under certain conditions with threaded couplings, in vertical risers, with special permission for branch circuit, feeder, or service mast installation and where support is provided by openings through framing members.

345-13 Boxes and Fittings

Refer to Article 370.

345-14 Splices and Taps

They are permitted only if made in accordance with Section 300-15. Refer to Article 370.

345-15 Bushings

A bushing must be used at the point where the conduit goes into a box, fitting, or other enclosure. This is to protect the wire. An exception is made if protection is provided with the box or fitting.

C. Construction Specifications

345-16 General

This type of conduit must be shipped in lengths of 10 ft. with one coupling. Smaller lengths can be shipped but only for a special use. If the conduit is nonferrous or corrosion-resistant material, it must be marked to indicate it. The conduit must be marked every 2½ ft. with the letters IMC.

ARTICLE 346
RIGID METAL CONDUIT

346-1 Use

Listed rigid metal conduit can be used in all locations and under all conditions, with the following provisos:

(a) If the conduit is steel and enamel coated, it can only be used indoors and not in highly corrosive areas.

(**b**) With the exception of aluminum fittings and enclosures and steel conduit, dissimilar metals in contact must not be used.

(**c**) It can be used in concrete, in contact with the earth, or in a corrosive area if properly protected and if it is suitable for that aim.

346-2 Other Articles

It must be in accordance with Article 300.

A. Installation

346-3 Cinder Fill

It cannot be used in or under cinder fill when there may be permanent moisture unless it is suitable for this type of installation, there are 2 in. or more of noncinder concrete all around it, it is at least 18 in. under the fill, or where suitable for the condition and has corrosion protection.

346-4 Wet Locations

Everything must be corrosion-resistant (bolts, straps, etc.).

346-5 Size

One-half inch is the minimum size permitted unless used for enclosing motor leads [see Section 430-145(b)]. The maximum size permitted is 6 inch.

346-6 Number of Conductors in Conduit

Refer to Chapter 9, Table 1.

346-7 Reaming and Threading

All rough edges must be removed when the conduit is cut. When threaded in the field, a tapered cutting die (¾ in. per foot taper) must be used.

346-8 Bushings

A bushing must be used at the point where the conduit goes into a box, fitting, or other enclosure. This is to protect the wire. An exception is made if protection is provided with the box or fitting.

346-9 Couplings and Connectors

Threadless couplings and connectors can be used. Running threads cannot be used. Use concrete type in concrete and raintight in wet areas.

346-10 Bends—How Made

The conduit must not be injured. Refer to Table 346-10 for the radius of the inner edge for field bends. If the conduit does not have a lead sheath and the bend is made with a bending machine that uses one operation, the Table 346-10 Exception can be used.

346-11 Bends—Number in One Run

The maximum that is permitted is not more than 360° in bends between pull points.

346-12 Supports

The conduit must be supported at intervals no longer than 10 ft. and fastened not more than 3 ft. from a box, fitting, etc. Fastening can be increased to 5 ft. where structural members do not permit fastening within 3 ft. There are four exceptions. An exception is made for straight runs with threaded couplings (refer to Table 346-12, page 267 in the *NEC*®) and no stress is transmitted to the terminations. Another exception is made for vertical risers from items like machine tools, where the distance between supports can be increased to 20 ft. if the conduit uses threaded couplings and is properly supported at the top and bottom. Exception 3 allows, where approved, that the conduit does not have to be fastened within 3 ft. of a service head for a termination of a mast above the roof. The last exception permits horizontal runs which are supported by openings through framing members not less than 10 ft. and fastened within 3 ft. of termination points.

346-13 Boxes and Fittings

Refer to Article 370.

346-14 Splices and Taps

They are permitted only in boxes or conduit bodies. Refer to Article 370 and Section 300-15.

TABLE 346-10 Radius of Conduit Bends (Inches)

Size of Conduit (In.)	Conductors without Lead Sheath (In.)	Conductors with Lead Sheath (In.)
½	4	6
¾	5	8
1	6	11
1¼	8	14
1½	10	16
2	12	21
2½	15	25
3	18	31
3½	21	36
4	24	40
4½	27	45
5	30	50
6	36	61

For SI units: (radius) 1 inch = 25.4 millimeters.

TABLE 346-10 Exception Radius of Conduit Bends (Inches)

Size of Conduit (In.)	Radius to Center of Conduit (In.)
½	4
¾	4½
1	5¾
1¼	7¼
1½	8¼
2	9½
2½	10½
3	13
3½	15
4	16
5	24
6	30

For SI units: (radius) 1 inch = 25.4 millimeters.

B. Construction Specifications

346-15 General

This type of conduit must be shipped in lengths of 10 ft. with one coupling. Smaller lengths can be shipped but only for special use. If the conduit is nonferrous of corrosion-resistant material, it must be marked to indicate it. At every 10 ft. the conduit must be marked with identification (refer to Section 110-21).

ARTICLE 347
RIGID NONMETALLIC CONDUIT

347-1 Description

This type of conduit is made from a nonmetallic material. It has to be able to withstand the load and impact in the installation being used (e.g., aboveground, underground, etc.). In addition, it must be moisture-resistant to chemical atmospheres and low temperature and sunlight, and flame-retardant if used aboveground.

347-2 Uses Permitted

Some of the uses permitted and other references in the *NEC*® are: in walls, floors, ceilings, and cinder fill, exposed when not subject to injury and approved; in corrosive locations where approved for that type of location and in accordance with Section 300-6; in dry and damp locations in accordance with Section 347-3; in wet locations where the material is approved; and underground in accordance with Sections 300-5 and 710-3(b).

347-3 Uses Not Permitted

This type of conduit must not be used in hazardous locations except where permitted by Sections 503-3(a), 504-20, 514-8, 515-5, and the Class 1, Division 2 locations in the Exception to Section 501-4(b); in areas where there might be physical damage to the conduit; in areas where the ambient temperature or the insulation temperature will be greater than allowed for the conduit and for supporting fixtures or equipment. There is an exception where it can be used to support nonmetallic conduit bodies.

347-4 Other Articles

It must be in accordance with Article 300. If equipment grounding is necessary (Article 250), a separate wire must be installed in the conduit. An exception for dc is noted in Section 250-57(b), Exception No. 2 and Section 250-57(b) Exception No. 3 for separately run equipment grounding conductors.

A. Installation

347-5 Trimming

Remove rough edges on cuts.

347-6 Joints

Use approved methods.

347-8 Supports

Refer to Section 300-18 and Table 347-8 for spacing of supports. There must be a support no farther than 3 ft. from a termination point. If it is listed for other securing and support spacing, it is permitted to follow but not exceed the listing. There is an exception for supports by openings through framing members.

347-9 Expansion Fittings

Expansion fittings have to be provided to compensate for thermal expansion and contraction where the change in length according to the table in Chapter 9 will be ¼ in. or greater between securely mounted conduit terminations.

347-10 Minimum Size—½ Inch

347-11 Number of Conductors

Refer to Chapter 9, Table 1.

347-12 Bushings

A bushing must be used at the point where the conduit goes into a box or fitting unless protection is provided in the box or fitting.

347-13 Bends—How Made

The conduit must not be injured. Refer to Table 346-10.

347-14 Bends—Number in One Run

The maximum that is permitted is not more than 360° in bends between pull points.

347-15 Boxes and Fittings

Refer to Article 370.

347-16 Splices and Taps

They are permitted only in accordance with Section 300-15. Refer to Article 370 for rules on installation of boxes and conduit bodies.

B. Construction Specifications

347-17 General

The conduit must be marked with identification every 10 ft. (refer to Section 110-21).

ARTICLE 348
ELECTRICAL METALLIC TUBING

348-1 Use

Listed electrical metal tubing (EMT) can be used in exposed or concealed areas. It cannot be used where it will be subject to severe damage, where protected from corrosion by only enamel, in cinder concrete or cinder fill where there is permanent moisture unless there is a layer of noncinder concrete on all sides at least 2 in. thick or the tubing is at least 18 in. under the fill, to support fixtures or equipment except conduit bodies not larger than the largest trade size of the tubing, or in hazardous locations except as permitted in Sections 502-4, 503-3, and 504-2. Where possible, there should be no contact between dissimilar metals in the system.

348-2 Other Articles

It must be in accordance with Article 300.

A. Installation

348-4 Wet Locations

Everything must be corrosion-resistant (bolts, straps, etc.).

348-5 Size

The minimum size permitted is ½ in. unless used for enclosing motor leads [see Section 430-145(b)]. The maximum size permitted is 4 in.

348-6 Number of Conductors in Tubing

Refer to Chapter 9, Table 1.

348-7 Threads

Threading is not permitted except on factory-threaded integral couplings.

348-8 Couplings and Connectors

They must be tight and concrete-tight if used in concrete and raintight if used in wet locations.

348-9 Bends—How Made

The tubing must not be injured. Refer to Table 346-10 for the radius of the inner edge of a field bend. When using a machine in the field designed for bending, Table 346-10 Exception can be used.

348-10 Bends—Number in One Run

The maximum that is permitted is not more than 360° inbends between pull points.

348-11 Reaming

Remove rough edges of all cuts.

348-12 Supports

The tubing must be installed as a complete system as noted in Article 300. The tubing must be fastened at intervals no longer than 10 ft. and not more than 3 ft. from a box, fitting, etc. The 3-ft. rule may be extended to 5 ft. for unbroken lengths where structural members interfere at 3 ft. There is an exception for concealed work. EMT in horizontal runs supported by openings through framing nt more than 10 ft. and fastened within 3 ft. of termination is permitted.

348-13 Boxes and Fittings

Refer to Article 370.

348-14 Splices and Taps

They are to be made in accordance with Section 300-15. Refer to Article 370.

B. Construction Specifications

348-15 General

The tubing and its fittings must be circular in cross section. It must be finished so that it can be distinguished from rigid metal conduit after its installation. If threaded couplings are used (when permitted), they must not bend the tubing at the threads. It must be clearly and durably marked at least every 10 ft. in accordance with Section 110-21.

ARTICLE 349
FLEXIBLE METALLIC TUBING

349-1 Scope

This tubing is circular, flexible, metallic, and liquidtight without a nonmetallic jacket. It can be used where there is no possibility of physical damage.

349-2 Other Articles

It must be in accordance with Article 300 and Section 110-21.

349-3 Uses Permitted

This type of tubing can be used for branch circuits in dry locations, and in accessible locations if not subject to damage or concealed, in accessible locations, and not more than nominal 1000 V systems.

349-4 Uses Not Permitted

It cannot be used in hoistways, storage battery rooms, hazardous locations, embedded in concrete, or directly buried in the earth. Lengths larger than 6 ft. are not permitted.

B. Construction and Installation

349-10 Size

The minimum size permitted is ½ in. except that ⅜ in. can be used in accordance with Section 300-22(b) and (c) or part of an approved lighting fixture assembly in accordance with Section 410-67(c) and not more than 6 ft. in length. The maximum size permitted is ¾ in.

349-12 Number of Conductors

Refer to Chapter 9, Table 1, for the number of conductors permitted in ½- and ¾-in. tubing. Refer to Table 350-3 for the number of conductors permitted in ⅜-in. tubing.

349-16 Grounding

Refer to Section 250-91(b), Exception No. 1.

349-17 Splices and Taps

349-18 Fittings

349-20 Bends

If the tubing is bent infrequently after installation, refer to Table 349-20(a), in the *NEC*®. If the tubing is bent for installation, use Table 349-20(b), in the *NEC*®.

ARTICLE 350
FLEXIBLE METAL CONDUIT

350-1 Scope

350-2 Definition

This is a circular raceway of helically wound, formed, interlocked metal strip.

350-3 Other Articles

Refer to the applicable paragraphs in Article 300.

350-4 Uses Permitted

It must be listed and can be used in exposed and concealed locations.

350-5 Uses Not Permitted

It cannot be used in wet locations unless approved for this and the liquid does not get into the conduit, in hoistways unless permitted by Section 620-21, in storage battery rooms, in hazardous locations unless allowed by Sections 501-4(b) and 504-20, where materials might deteriorate conductors, underground or in poured concrete or aggregate, and where subject to physical damage.

B. Installation

350-10 Size

Minimum size permitted is ½ in. The use of ⅜ in is allowed for leads to motors [Section 430-145(b)], not longer than 6 ft. as part of a listed assembly or taps to lighting fixtures [Section 410-67(c)] or utilization equipment, for manufactured wiring systems [Section 604-6(a)], in hoistways Section 620-21, and as permitted by Section 410-77(c) as part of a listed assembly to connect wired fixture sections. There are five exceptions. The maximum size permitted is 4 in.

350-12 Number of Conductors

Cannot exceed the percentage fill of Table 1, Chapter 9 or Table 350-12 for ⅜ in.

350-14 Grounding

This type of conduit can be used as a grounding means as noted in Section 250-91(b). Refer to the *NEC*® for exceptions and bonding jumpers.

350-16 Bends

The maximum number that is permitted is not more than 360 degrees in bends between pull points. The radius of curve of the inner edge of a field bend must comply with Table 346-10.

350-18 Supports

Securely fasten within 12 in. of a termination point and supports must be not more than 4½ ft. apart. There is a provision for supports through framing openings. There are three exceptions.

350-20 Fittings

Use only listed fittings.

350-22 Trimming

350-23 Splices and Taps

They can only be made in boxes and conduit bodies. Refer to Article 370.

ARTICLE 351
LIQUIDTIGHT FLEXIBLE METAL CONDUIT AND LIQUIDTIGHT FLEXIBLE NONMETALLIC CONDUIT

351-1 Scope

A. Liquidtight Flexible Metal Conduit

351-2 Definition

This conduit is similar to flexible metal conduit, but it has a jacket over the metal conduit. The jacket is liquidtight, nonmetallic, and sunlight-resistant.

351-3 Other Articles

It must be installed in accordance with Articles 300, 350, 501, 502, 503, and 553, where applicable.

351-4 Use

This conduit can be installed exposed or concealed, directly buried in earth and if protection from liquids, vapor, or solids is necessary or if flexibility is desired. It can be installed in hazardous areas if approved for that installation. Refer to Sections 501-4(b), 502-4, 503-3, 504-20 and 553-7(b). It cannot be installed when subject to damage or where temperatures exceed that which is approved for the material.

351-5 Size

Minimum ½ in. For exceptions, refer to Section 350-10(a). The maximum size allowed is 4 in.

351-6 Number of Conductors

Refer to Chapter 9, Table 1. For ⅜-in. conduit refer to Table 350-3 in the *NEC*®.

351-7 Fittings

351-8 Supports

It must be supported and secured not more than 4½ ft. apart. It must be securely fastened within 12 in. of a conduit termination. Three exceptions are for when the conduit is fished, where flexibility is needed, length not exceeding 3 ft. at terminals, and lengths not exceeding 6 ft. from a fixture terminal connection for tap conductors to a lighting fixture as noted in Section 410-67(c). Horizontal runs allowed to be supported by the openings in framing members not more than 4½ ft. apart and fastened within 12 in. of a termination.

351-9 Grounding

This type of conduit can be used as a means of grounding under certain conditions. Refer to the *NEC*®.

351-10 Bends

The maximum that is permitted is not more than 360° in bends between pull points.

351-11 Splices and Taps

They can only be made in boxes or conduit bodies in accordance with Section 300-15. Refer to Article 370 for information on use of boxes and conduit bodies.

B. Liquidtight Flexible Nonmetallic Conduit

351-22 Definition

This conduit has a circular cross section. It comes in various types. It can have a smooth inner surface with integral reinforcement; smooth seamless inner core and cover bonded together with the reinforcement between the core and cover; or corrugated surfaces without reinforcement within the walls. This type of conduit is flame resistant and is approved for electrical conductors, with fittings.

351-23 Use

This conduit can be installed exposed or concealed if protection from vapors, liquids, or solids is necessary or if flexibility is desired. It cannot be installed if subject to damage, when temperatures will exceed those for which the conduit is approved, in lengths in excess of 6 ft., and where voltages higher than 600 V are used. There are exceptions to the latter two conditions. It can also be used outdoors when listed and suitable and for direct burial use when listed and marked.

351-24 Size

½ to 4 in. For leads of motors ⅜ in. can be used in accordance with Section 430-145(b). There are two other exceptions.

351-25 Number of Conductors

Refer to Chapter 9, Table 1.

351-26 Fittings

351-27 Supports

Refer to Sectoin 351-22. There is an additional exception for lengths longer than 6 ft.

351-28 Equipment Grounding

If required, install a separate wire inside or outside of conduit. Refer to *NEC®* for restrictions.

351-29 Splices and Taps

They can only be made in boxes or conduit bodies in accordance with Section 300-15. Note Article 370.

351-30 Bends—Number in One Run

The maximum that is permitted is not more than 360° in bends between pull points.

ARTICLE 352
SURFACE METAL RACEWAYS AND SURFACE NONMETALLIC RACEWAYS

A. Surface Metal Raceways

352-1 Use

This can be installed in dry areas. It cannot be installed where it might be damaged, where there are corrosive vapors, in hoistways, where the voltage between conductors is 300 V or more unless the metal is at least 0.04 in., in a hazardous area [refer to Section 501-4(b) for an exception], or concealed. It can be installed where permitted by Section 645-2(c)(2).

352-2 Other Articles

It must be installed in accordance with Article 300.

352-3 Size of Conductors

They cannot be larger than what the raceway was designed for.

352-4 Number of Conductors in Raceways

Refer to design criteria. Exceptions to the use of derating factors are noted in the *NEC*® in this paragraph.

352-5 Extension Through Walls and Floors

This can be done if the material is dry and no joints are in the extension.

352-6 Combination Raceways

352-7 Splices and Taps

352-8 General

352-9 Grounding

B. Surface Nonmetallic Raceways

352-21 Description

This raceway is similar to the metallic type except that it is made of nonmetallic material that is resistant to moisture atmospheres, chemical atmospheres, impact and crushing, distortion from heat, and low temperatures. It is also flame-retardant. They can have the suffix LS.

352-22 Use

This can be used in dry areas. It cannot be installed concealed, where it might be damaged, where the voltage between conductors is 300 V or more unless so listed, in hoistways, in a hazardous area [refer to Section 501-4(b) for an exception], if ambient temperatures are higher than that for which the raceway is listed, or for conductors with insulation temperatures higher than the raceway is listed.

352-23 Other Articles

It must be installed in accordance with Article 300.

352-24 Size of Conductors

Refer to Section 352-3.

352-25 Number of Conductors in Raceways

This is the same as Section 352-4.

352-26 Combination Raceways

352-27 General

352-28 Extension Through Walls and Floors

352-29 Splices and Taps

C. Strut-Type Channel Raceway

352-40 Description

352-41 Uses Permitted

352-42 Uses Not Permitted

352-43 Other Articles

352-44 Size of Conductors

352-45 Number of Conductors in Raceways

352-46 Extension Through Walls and Floors

352-47 Support of Strut-Type Channel Raceways

352-48 Splices and Taps

352-49 General

352-50 Grounding

352-51 Marking

ARTICLE 353
MULTIOUTLET ASSEMBLY

353-1 Other Articles

353-2 Use

353-3 Metal Multioutlet Assembly through Dry Partitions

ARTICLE 354
UNDERFLOOR RACEWAYS

354-1 Other Articles

354-2 Use

This type of raceway can be used in an office in the concrete floor or other floorings. It cannot be used if exposed to corrosive vapors, in a hazardous area [for an exception refer to Sections 501-4(b) and 504-20], in concrete, or in corrosive areas unless properly protected and approved.

354-3 Covering

354-4 Size of Conductors

Not larger than designed for.

354-5 Maximum Number of Conductors in Raceway

Not more than 40% fill.

354-6 Splices and Taps

354-7 Discontinued Outlets

354-8 Laid in Straight Lines

354-9 Marker at Ends

354-10 Dead Ends

354-13 Junction Boxes

354-14 Inserts

354-15 Connections to Cabinets and Walls

ARTICLE 356
CELLULAR METAL FLOOR RACEWAYS

356-1 Definition

This is a raceway system that is part of the entire floor and installed as such. A complete definition of terms is given in the *NEC®*.

356-2 Uses Not Permitted

356-3 Other Articles

A. Installation

356-4 Size of Conductors

Not larger than 1/0 unless special permission is obtained.

356-5 Maximum Number of Conductors in Raceway

Not more than 40% fill.

356-6 Splices and Taps

356-7 Discontinued Outlets

356-8 Markers

356-9 Junction Boxes

356-10 Inserts

356-11 Connection to Cabinets and Extensions from Cells

B. Construction Specifications

356-12 General

The following article titles are listed for reference:

Article 358—Cellular Concrete Floor Raceways; **Article 362**—Metal Wireways and Nonmetallic Wireways; **Article 363**—Flat Cable Assemblies—Type FC; **Article 364**—Busways; **Article 365**—Cablebus.

ARTICLE 370
OUTLET, DEVICE, PULL AND JUNCTION BOXES, CONDUIT BODIES AND FITTINGS

A. Scope and General

370-1 Scope

Special requirements for installations over 600 V are indicated in Part D.

370-2 Round Boxes

These are not permitted if locknuts or bushings are required on the side of the box.

370-3 Nonmetallic Boxes

These are only to be used with nonmetallic cable or conduit, open wire on insulator installations, or knob-and-tube concealed wiring. They can be used with metal raceway or metal-jacketed cable only if there is a means for integral bonding or internal bonding.

370-4 Metal Boxes

These must be grounded as provided in Article 250.

370-5 Short Radius Conduit Bodies

Conduit bodies, like capped elbows, which enclose No. 6 or smaller conductors and are used only to allow the installation of the raceway and its conductors can not have splices, taps, or devices and they must be large enough for free space for the conductors.

B. Installation

370-15 Damp, Wet Locations, or Hazardous (Classified) Locations

Moisture must be prevented from getting into the box. They must be listed for use in wet locations. When installed in a hazardous location, conform to Articles 500 through 517.

370-16 Number of Conductors in Outlet, Device and Junction Boxes, and Conduit Bodies

This is a new section and the reader is referred to it for complete information. These items must be large enough to allow free space for all the enclosed conductors. The volume of the box as calculated in this section cannot be smaller than the fill calculation. This does not apply to terminal housings supplied with the motors (refer to Section 430-12). For enclosed conductors No. 4 or larger the provisions of Section 370-28 must be complied with.

(a) Box Volume Calculations. The volume is the total of the assembled sections and if used, spaces from plaster rings, domed covers, extension rings, and the like that are made from boxes listed in Table 370-16(a) or have the volume marked.

(1) Standard Boxes. Given in Table 370-16(a).

(2) Other Boxes. If the box is 100 cubic inches or less [except those in Table 370-16(a)] and nonmetallic boxes must have the capacity marked by the manufacturer. If the box is listed in Table 370-16(a) and has a larger capacity, the manufacturer can mark it.

TABLE 370-16(a) Metal Boxes

Box, Dimension, Inches Trade Size or Type	*Min. Cu. In. Cap.*	*Maximum Number of Conductors**						
		No. 18	*No. 16*	*No. 14*	*No. 12*	*No. 10*	*No. 8*	*No. 6*
4 × 1¼ round or octagonal	12.5	8	7	6	5	5	4	2
4 × 1½ round or octagonal	15.5	10	8	7	6	6	5	3
4 × 2⅛ round or octagonal	21.5	14	12	10	9	8	7	4
4 × 1¼ square	18.0	12	10	9	8	7	6	3
4 × 1½ square	21.0	14	12	10	9	8	7	4
4 × 2⅛ square	30.3	20	17	15	13	12	10	6
4¹¹⁄₁₆ × 1¼ square	25.5	17	14	12	11	10	8	5
4¹¹⁄₁₆ × 1½ square	29.5	19	16	14	13	11	9	5
4¹¹⁄₁₆ × 2⅛ square	42.0	28	24	21	18	16	14	8
3 × 2 × 1½ device	7.5	5	4	3	3	3	2	1
3 × 2 × 2 device	10.0	6	5	5	4	4	3	2
3 × 2 × 2¼ device	10.5	7	6	5	4	4	3	2

3 × 2 × 2½ device	12.5	8	7	6	5	5	4	2
3 × 2 × 2¾ device	14.0	9	8	7	6	5	4	2
3 × 2 × 3½ device	18.0	12	10	9	8	7	6	3
4 × 2⅛ × 1½ device	10.3	6	5	5	4	4	3	2
4 × 2⅛ × 1⅞ device	13.0	8	7	6	5	5	4	2
4 × 2⅛ × 2⅛ device	14.5	9	8	7	6	5	4	2
3¾ × 2 × 2½ masonry box/gang	14.0	9	8	7	6	5	4	2
3¾ × 2 × 3½ masonry box/gang	21.0	14	12	10	9	8	7	4
FS—minimum internal depth 1¾ single cover/gang	13.5	9	7	6	6	5	4	2
FD—minimum internal depth 2⅜ single cover/gang	18.0	12	10	9	8	7	6	3
FS—minimum internal depth 1¾ multiple cover/gang	18.0	12	10	9	8	7	6	3
FD—minimum internal depth 2⅜ multiple cover/gang	24.0	16	13	12	10	9	8	4

For SI units: one cubic inch = 16.4 cm^3.

* Where no volume allowances are required by Sections 370-16(b)(2) through 370-16(b)(5).

(b) Box Fill Calculations.

(1) Conductor Fill. Every conductor that originates outside the box and terminates or is spliced in the box and each conductor passing through the box is counted once. The fill is computed using Table 370-16(b). If a conductor has no part leaving the box it is not counted. There is an exception for equipment grounding conductors.

(2) Clamp Fill. If an internal clamp is used there, a single volume allowance in accordance with Table 370-16(b) is made based on the largest conductor in the box.

(3) Support Fittings Fill. If a stud or hickey is there a single volume allowance in accordance with Table 370-16(b) is made based on the largest conductor in the box.

(4) Device or Equipment Fill. For a strap or yoke containing a device or equipment a double volume allowance is made in accordance with Table 370-16(b) based on the largest conductor connected to the device or equipment.

(5) Equipment Grounding Fill. If equipment grounding conductors come into the box, a single volume allowance is made in accordance with Table 370-16(b) and is based on the largest equipment grounding conductor present. If additional equipment grounding conductors as noted in Section 250-74, Exception No. 4 come into the box an additional volume allowance is made based on the largest equipment grounding conductor in this set.

(c) Conduit Bodies. Generally they must have cross-sectional area not less than twice the area of the largest conduit or tubing attached. The maximum number of conductors is shown in Table 1, Chapter 9. There are additional requirements noted in the Code.

370-17 Conductors Entering Boxes, Conduit Bodies, or Fittings

All openings must be closed. All conductors coming into a box, conduit body or fitting must be protected against abrasion. Special requirements are indicated with open wiring or concealed knob-and-tube wiring. If cable or raceway is used the cable or

TABLE 370-16(b) Volume Required per Conductor

Size of Conductor	*Free Space within Box for Each Conductor*
No. 18	1.5 cubic inches
No. 16	1.75 cubic inches
No. 14	2. cubic inches
No. 12	2.25 cubic inches
No. 10	2.5 cubic inches
No. 8	3. cubic inches
No. 6	5. cubic inches

For SI units: one cubic inch = 16.4 cm^3.

raceway must be fastened to the box, conduit body or fitting. Refer to Section 370-17 of the *NEC®*, for all rules and to Section 373-6(c) for wire No. 4 or larger.

370-18 Unused Openings

These openings must be closed. When metal plugs are used in nonmetallic boxes, they must be recessed at least ¼ in.

370-19 Boxes Enclosing Flush Devices

They must be closed on the back and sides and a support for the device (usually a stud or screw hole) must be in the box. The screw that supports the box cannot be used to attach the device.

370-20 In Wall or Ceiling

When the box is set in concrete, tile, or other noncombustible material, it can be set back as much as ¼ in. When it is set in combustible material, it has to be flush with the finished surface.

370-21 Repairing Plaster and Drywall or Plasterboard

370-22 Exposed Surface Extensions

370-23 Supports

This section details methods of support. It indicates where they are needed and what can be used.

370-24 Depth of Outlet Boxes

370-25 Covers and Canopies

370-27 Outlet Boxes

They must be designed for the purpose to be used (i.e., floor or lighting fixture). A fixture support must be in a lighting fixture box. Cannot be used as sole support for ceiling paddle fans unless listed for that purpose.

370-28 Pull and Junction Boxes

(a) Minimum Size. The following rules apply only when the wire is No. 4 or larger. In a straight pull, the box must be at least 8 times the trade diameter of the largest raceway in length. When angle pulls or U-pulls are used, the distance between the raceway entrances and the opposite wall must be at least 6 times the trade diameter of the largest raceway in a row. In addition, that distance must be increased by the diameter of each additional raceway in any one row on the same wall of the box. Since each row is calculated individually, the largest distance for a single row is used. Two raceways that enclose the same conductor cannot be closer than 6 times the trade diameter of the larger raceway. There is an exception where the entry is opposite a removable cover. There is also a provision for boxes or conduit bodies whose dimensions are less than calculated where combination of conductors are used.

(b) Conductors in Pull or Junction Boxes.

(c) Covers.

(d) Permanent Barriers. If a permanent barrier is in a box, each section is considered a box.

370-29 Conduit Bodies, Junction, Pull, and Outlet Boxes to Be Accessible

C. Construction Specifications

370-40 Metal Boxes, Conduit Bodies, and Fittings

370-41 Covers

370-42 Bushings

370-43 Nonmetallic Boxes

370-44 Marking

D. Pull and Junction Boxes for Use on Systems Over 600 Volts, Nominal

370-70 General

370-71 Size of Pull and Junction Boxes

370-72 Construction and Installation Requirements

ARTICLE 373
CABINETS, CUTOUT BOXES, AND METAL SOCKET ENCLOSURES

373-1 Scope

A. Installation

373-2 Damp, Wet, or Hazardous (Classified) Locations

When installed in wet locations, they must be weatherproof. In damp or wet locations, surface boxes must prevent moisture from entering and have at least ¼ in. of air space between it and the wall or support. Nonmetallic cabinets and cutout boxes can be installed on concrete, tile, or masonry without an airspace. In hazardous locations conformance must be made with the requirements of Articles 500 through 517.

373-3 Position in Wall

When installed in noncombustible material, they may be set back ¼ in. When installed in combustible material, they must be flush with the surface.

373-4 Unused Openings

These must be closed. When metal plugs are used in nonmetallic boxes, they must be recessed at least ¼ in.

373-5 Conductors Entering Cabinets or Cutout Boxes

The opening must be closed. When metal cabinets or boxes are used with open wiring or knob and tube work, the conductor must come in through a bushing. In dry areas, it may enter through flexible tubing. Cables must be secured to the box.

373-6 Deflection of Conductors

Refer to Section 373-6 and Tables 373-6(a) and 373-6(b) in the *NEC*® for minimum width of gutters and bending spaces.

373-7 Space in Enclosure

373-8 Enclosures for Switches or Overcurrent Devices

373-9 Side or Back Wiring Spaces or Gutters

B. Construction Specifications

373-10 Material

If metal, it must have corrosion protection. Sheet steel cabinets must be at least 0.053 in. thick uncoated. Nonmetallic cabinets must be approved by the authorities.

373-11 Spacing

ARTICLE 374
AUXILIARY GUTTERS

374-1 Use

374-2 Extension Beyond Equipment

It must not extend more than 30 ft. from the equipment it supplements.

374-3 Supports

374-4 Covers

374-5 Number of Conductors

Not more than 30 conductors are allowed at any point and the area of the conductors cannot be more than 20% of the area of the gutter. There are three exceptions to this: (1) for elevators; (2) for signaling or motor controller circuits; and (3) when the derating factors are used for ampacity of conductors, the 30-conductor limit does not apply.

374-6 Ampacity of Conductors

The following refer to sheet metal auxiliary gutters. When there are not more than 30 current-carrying conductors, the derating factors for ampacities do not have to be applied. The capacities of bars cannot exceed 1000 A per square inch for copper and 700 A per square inch for aluminum. For nonmetallic auxiliary gutters use the derating factors in Article 310 Note 8(a) of Notes to Ampacity Tables of 0 to 2000 Volts.

374-7 Clearance of Bare Live Parts

374-8 Splices and Taps

374-9 Construction and Installation

ARTICLE 380 SWITCHES

380-1 Scope

380-2 Switch Connections

Three- and four-way switches must be installed so that switching is in the underground conductor. If the wiring is installed in metal enclosures or metal-jacketed cable, it must be in accordance with the requirements of Section 300-20(a). A grounded conductor must not be switched. There are two exceptions to this.

380-3 Enclosure

380-4 Wet Locations

380-5 Time Switches, Flashers, and Similar Devices

380-6 Position of Knife Switches

380-7 Indicating

380-8 Accessibility and Grouping

380-9 Faceplates for Flush-Mounted Snap Switches

380-10 Mounting of Snap Switches

380-11 Circuit Breakers at Switches

380-12 Grounding of Enclosures

380-13 Knife Switches

380-14 Rating and Use of Snap Switches

B. Construction Specifications

380-15 Marking

They must be marked with current, voltage, and the maximum horsepower if horsepower rated.

380-16 600-Volt Knife Switches

380-17 Fused Switches

Parallel fusing is not permitted except as noted in Section 240-8, Exception.

380-18 Wire Bending Space

ARTICLE 384
SWITCHBOARDS AND PANELBOARDS

A. General

384-1 General

Switchboards, panelboards, distribution boards, and battery-charging panels are covered.

384-2 Other Articles

Articles 240, 250, 370, 373, and 380 and applicable sections of other Code articles must be complied with.

384-3 Support and Arrangement of Busbars and Conductors

They must be located to be free from damage and be secured. They must be arranged to avoid overheating due to induction. When used as service equipment, a grounding bar must be provided and it must be bonded to the case and frame. There is an exception for high-impedance grounded neutral system connections. Locations of terminals must be so as to avoid the necessity of reaching across ungrounded line bus. In a four-wire delta system the bar or conductor with the highest voltage to ground must be marked. Three-phase buses must have the phases A, B, C, front to back, left to right, or top to bottom when looking from the front. There is an exception for equipment in the same switchboard or panelboard as the meter. Refer to Section 373-6 for minimum wire bending space.

384-4 Installation

No piping, ducts, equipment, etc., can pass through the room or space designated for a switchboard or panelboard. Clear space requirements are noted.

B. Switchboards

384-5 Location of Switchboards

384-6 Switchboards in Damp or Wet Locations

384-7 Location Relative to Easily Ignitable Material

384-8 Clearances

At least 3 ft. to a combustible ceiling. Refer to Section 110-16 for clearances around the switchboard.

384-9 Conductor Insulation

384-10 Clearance for Conductors Entering Bus Enclosure

384-11 Grounding Switchboard Frames

They must be grounded. There is an exception for two-wire dc switchboards that are effectively insulated from ground.

384-12 Grounding of Instruments, Relays, Meters, and Instrument Transformers on Switchboards

Refer to Sections 250-121 through 250-125.

C. Panelboards

384-13 General

They must have current, voltage, phase, and manufacturer's trademark markings visible after installation. The rating must not be less than the feeder capacity. All circuits and modifications must be noted on a directory on the panelboard.

384-14 Lighting and Appliance Branch-Circuit Panelboard

If a panelboard has more than 10% of its overcurrent devices rated less than 30 A with a neutral connection, it falls into this category.

384-15 Number of Overcurrent Devices on One Panelboard

Not more than 42 overcurrent devices can be installed in a lighting or appliance branch-circuit panelboard. This does not include mains. Each pole of a multipole circuit breaker is considered a single circuit.

384-16 Overcurrent Protection

(a) Individual overcurrent protection must be provided for a lighting and appliance branch-circuit panelboard of not more than two sets of circuit breakers or fuses. The overcurrent protection cannot exceed the ratings of the panelboard. This is not required if the feeder has overcurrent protection that does not exceed the rating of the panelboard. If an existing panelboard is used as service equipment in a single-residence occupancy, the individual protection is not required.

(b) Overcurrent protection not greater than 200 A is required for a panelboard that has snap switches rated 30 A or less.

(c) Where the load will continue for at least 3 hours on an overcurrent device in the panelboard, the load cannot be greater than 80% of the device's rating. This is not required if the assembly and the device are rated for 100% continuous duty.

(d) For panelboards that are supplied through a transformer, the overcurrent protection required must be on the secondary of the transformer. This is not required for two-wire secondaries (single voltage). Refer to Section 450-3(b)(1).

(e) A panelboard with less than three-phase buses cannot use a three-phase overcurrent device or disconnect.

(f) An additional means of fastening that requires a release other than a pull to release the device must be used for backfeed plug-in devices (plug-in overcurrent or plug-in main lug assemblies).

384-17 Panelboards in Damp or Wet Locations

Refer to Section 373-2(a).

384-18 Enclosure

Panelboards must be in some type of enclosure with a dead front. If only qualified persons have access to the panelboards, dead fronts are not required.

384-19 Relative Arrangement of Switches and Fuses

Fuses must be installed on the load side of switches.

384-20 Grounding of Panelboards

Metal cabinets and frames must be in contact with each other and grounded in accordance with Article 250 or Section 384-3(c). A terminal bar for grounding conductors must be inside the cabinet when separate grounding conductors or nonmetallic raceway or cable is used. If the panelboard frame and cabinet are metal, the terminal bar is bonded to them. If nonmetallic, the terminal bar is connected to the grounding conductor from the feeder. The terminal bar used for the grounding conductors must be identified and located where the connection is made from the grounded conductor to a grounding electrode in accordance with Article 250. An isolated ground conductor can pass through the panelboard with connection to the ground bar in accordance with Section 250-74, Exception No. 4.

D. Construction Specifications

384-30 Panels

Must be made of moisture-resistant noncombustible material for switchboards.

384-31 Busbars

Must be rigidly mounted.

384-32 Protection of Instrument Circuits

Instruments must be protected and the rating must not exceed 15 A.

384-33 Component Parts

They must be in accordance with Articles 240 and 380.

384-34 Knife Switches

384-35 Wire Bending Space in Panelboards

Top and bottom wire bending space must be in accordance with Table 373-6(b). Side bending space must be in accordance with Table 373-6(a). Refer to this article in the *NEC*® for exceptions.

384-36 Minimum Spacings

Refer to Table 384-36.

4

EQUIPMENT FOR GENERAL USE

ARTICLE 400
FLEXIBLE CORDS AND CABLES

A. General

400-1 Scope

400-2 Other Articles

400-3 Suitability

400-4 Types

Table 400-4 lists the approved types of flexible cables and cords. A cable or cord not on this list must have a special investigation. Notes to the table should be reviewed.

400-5 Ampacity of Flexible Cords and Cables

Tables 405-5(A) and 405-5(B) show the ampacity of flexible cords or cables of up to three current-carrying conductors. When more than three current-carrying conductors are used, a derating factor is used. These derating factors are shown in a table in the code. There is information on Ultimate Insulation Temperature.

400-6 Marking

400-7 Uses Permitted

(1) Pendants; (2) wiring of fixtures; (3) connection of portable lamps or appliances; (4) for elevator cables; (5) cranes and hoists; (6) for stationary equipment if they are to be moved a lot; (7) to prevent noise and vibration transmission; (8) where removal for maintenance is required and the appliance is to use a flexible cord or cable; (9) for data processing cables (refer to Section 645-15); (10) connection of moving parts; (11) as allowed in Sections 305-4(b) and 305-4(c) for the connection of moving parts. When used as indicated in (3), (6), and (8), a plug must be used on the cord and it must be plugged into a receptacle for power.

400-8 Uses Not Permitted

Unless permitted in Section 400-7, they are not permitted in: (1) Fixed wiring of a building; (2) installed through a hole in walls, ceilings, or floors; (3) through doorways, windows, or similar openings; (4) installed concealed; (5) attached to building surfaces; (6) installed in raceways unless permitted by the code. There is an exception for a tension take-up device.

400-9 Splices

400-10 Pull at Joints and Terminals

Some type of strain relief must be provided (i.e., a knot, tape, or special fitting).

400-11 In Show Windows and Show Cases

Use only Type S, SO, SE, SEO, SEOO, SOO, SJ, SJE, SJO, SJEO, SJEOO, SJOO, ST, STO, STOO, SJT, SJTO, SJTOO, or AFS. This restriction is not necessary for chain-supported lighting fixtures or portable lamps and display merchandise.

400-12 Minimum Size

Refer to Table 400-4.

400-13 Overcurrent Protection

If not smaller than No. 18 or tinsel cords or a cord approved for a specific appliance, no separate overcurrent protection is required.

400-14 Protection from Damage

B. Construction Specifications

400-20 Labels

400-21 Nominal Insulation Thickness

Refer to Table 400-4.

400-22 Grounded-Conductor Identification

One conductor that is used as a grounded circuit conductor must be clearly marked. Identification methods include (1) a white or gray braid, (2) a tracer of a distinguishable color in the braid, (3) white or gray insulation, (4) white or gray separation, (5) tinned conductor, and (6) some distinguishable surface marking.

400-23 Equipment Grounding-Conductor Identification

This conductor must also have a distinguishing marking of a colored braid, colored insulation, or colored cover. The color to be used is either green or green with yellow stripes.

400-24 Attachment Plugs

C. Portable Cables Over 600 Volts, Nominal

400-30 Scope

400-31 Construction

400-32 Shielding

400-33 Grounding

400-34 Minimum Bending Radii

400-35 Fittings

400-36 Splices and Terminations

ARTICLE 402
FIXTURE WIRES

402-1 Scope

402-2 Other Articles

402-3 Types

Refer to Table 402-3 for those types permitted. They can be used at 600 V.

402-5 Allowable Ampacities for Fixture Wires

Table 402-5 specifies the ampacity of fixture wire.

402-6 Minimum Size

Not smaller than No. 18.

402-7 Number of Conductors in Conduit or Tubing

Refer to Chapter 9, Table 2.

402-8 Grounded-Conductor Identification

402-9 Marking

402-10 Uses Permitted

Lighting fixtures and similar equipment where enclosed and protected and for connecting lighting fixtures to branch circuits.

402-11 Uses Not Permitted

Cannot be used as branch-circuit conductors except in accordance with Sections 725-27 and 760-27.

402-12 Overcurrent Protection

ARTICLE 410 LIGHTING FIXTURES, LAMPHOLDERS, LAMPS, AND RECEPTACLES

A. General

410-1 Scope

410-2 Application of Other Articles

410-3 Live Parts

B. Fixture Locations

410-4 Fixtures in Specific Locations

(a) Wet and Damp Locations. The fixtures must be installed in a manner so that water will not enter or collect in the lampholder, wireway, etc. Examples of wet locations are underground, in concrete slabs, where it can be saturated with water, etc. Examples of damp locations are partially protected areas such as under canopies, roofs, or interiors, where subject to moisture, as in farms.

(b) Corrosive Locations

(c) In Ducts or Hoods. This location is permitted only in nonresidential cooking areas and is subject to four specific conditions.

(d) Above Bathtubs

410-5 Fixtures near Combustible Material

The fixture construction must ensure that the combustible material will not be subjected to temperatures over 194°F.

410-6 Fixtures over Combustible Material

These must be unswitched fixtures. The lampholder must be at least 8 ft. above the floor unless an individual switch is installed for each fixture.

410-7 Fixtures in Show Windows

External wiring is not permitted except chain-supported fixtures.

410-8 Fixtures in Clothes Closets

(a) Definition. The code has a specific set of dimensions for storage space with an explanatory diagram.

(b) Fixture Types Permitted. Surface-mounted or recessed incandescent or fluorescent fixtures are permitted. The incandescent fixture must have a completely enclosed lamp.

(c) Fixtures Types Not Permitted. Pendant fixtures or lampholders and incandescent fixtures whose lamps are not totally enclosed.

(d) Location. Surface-mounted fixtures can be installed on the wall above the door or on the ceiling. The clearance between the fixture and nearest point of storage must be 12 in. for an incandescent fixture and 6 in. for a fluorescent fixture. Recessed fixtures can be installed in the wall or ceiling. The clearance between the fixture and nearest point of storage must be 6 in.

410-9 Space for Cove Lighting

C. Provisions at Fixture Outlet Boxes, Canopies, and Pans

410-10 Space for Conductors

410-11 Temperature Limit of Conductors in Outlet Boxes

410-12 Outlet Boxes to be Covered

410-13 Covering of Combustible Material at Outlet Boxes

410-14 Connection of Electric-Discharge Lighting Fixtures

D. Fixture Supports

410-15 Supports

410-16 Means of Support

The *NEC*® specifies a few requirements for outlet boxes, inspection, suspended ceilings, fixture studs, insulating joints, raceway fittings, busways, and trees. One important requirement to note is that a fixture that weighs more than 50 lb cannot be supported from an outlet box. It must be independently supported.

E. Grounding

410-17 General

410-18 Exposed Fixture Parts

If the wiring system provides a means of grounding, exposed conductive parts must be grounded. If no provision for grounding is available, they must be insulated.

410-19 Equipment Over 150 Volts to Ground

All metal fixtures, transformers, and transformer enclosures must be grounded. All other metal parts must be insulated or grounded.

410-20 Equipment Grounding Conductor Attachment

A means to connect an equipment ground must be provided for a metal fixture.

410-21 Methods of Grounding

Mechanical connection to an equipment grounding conductor in accordance with Sections 250-91(b) and 250-95.

F. Wiring of Fixtures

410-22 Fixture Wiring—General

410-23 Polarization of Fixtures

410-24 Conductors

410-25 Conductors for Certain Conditions

410-27 Pendant Conductors for Incandescent Filament Lamps

410-28 Protection of Conductors and Insulation

410-29 Cord-Connected Showcases

410-30 Cord-Connected Lampholders and Fixtures

410-31 Fixtures as Raceways

Normally, fixtures cannot be used as raceways. It is permitted, however, if the fixture is listed for that use, when the fixtures are designed for end-to-end assembly. Only two two-wire circuits are permitted. If the branch-circuit conductors come within 3 in. of a ballast within the ballast compartment, the wire must have a usage rating of at least 194°F.

G. Construction of Fixtures

410-34 Combustible Shades and Enclosures

410-35 Fixture Rating

410-36 Design and Material

410-37 Nonmetallic Fixtures

410-38 Mechanical Strength

410-39 Wiring Space

410-42 Portable Lamps

410-44 Cord Bushings

410-45 Tests

410-46 Live Parts

H. Installation of Lampholders

410-47 Screw-Shell Type

410-48 Double-Pole Switched Lampholders

410-49 Lampholders in Wet or Damp Locations

J. Construction of Lampholders

410-50 Insulation

410-51 Lead Wires

410-52 Switched Lampholders

K. Lamps and Auxiliary Equipment

410-53 Bases, Incandescent Lamps

410-54 Electric-Discharge Lamp Auxiliary Equipment

L. Receptacles, Cord Connectors, and Attachment Plugs (Caps)

410-56 Rating and Type

Receptacles cannot be used as lampholders and must be rated at least 15 A 125 V or 15 A 250 V. This section also specifies requirements for CO/ALR receptacles, isolated ground receptacles, faceplates, position of receptacle faces, attachment plugs, attachment plug ejector mechanisms, noninterchangeability, and receptacles in raised covers. One thing to note is that the receptacle must project at least 0.015 in. from a metal faceplate.

410-57 Receptacles in Damp or Wet Locations

Receptacles located in damp areas must have a weatherproof enclosure. The receptacle need only by protected when it is covered, not when it is in use. Receptacles located in wet locations must be weatherproof-protected at all times, even when in use. An exception to the wet location rule is made only if portable tools or plug-in devices are used that will not be left plugged in when not in use. In these cases, the protection provided for damp locations is acceptable. Receptacles located outdoors must be installed high enough so that water will not touch the plate or receptacle (snow or water accumulation). Floor receptacles must be installed so as not to be damaged by cleaning equipment. If the receptacle is flush-mounted, the faceplate must provide the weatherproofing (gasketing, etc.). Receptacles cannot be installed within a bathtub or shower space.

410-58 Grounding-Type Receptacles, Adapters, Cord Connectors, and Attachment Plugs

Grounding-type receptacles must have a grounding slot. Grounding-type plugs must have a grounding prong and this prong must make contact first. The ground pole must be distinctively marked and it must not be used for any other purpose.

M. Special Provisions for Flush and Recessed Fixtures

410-64 General

This type of fixture must comply with Sections 410-65 through 410-72 as well as other parts of the *NEC*®.

410-65 Temperature

They must be installed so that any adjacent combustible material will not be subjected to a temperature higher than 194°F. If the fixture is recessed in fire-resistant construction in a fire-resistant building, the temperature restriction is increased to 302°F if the fixture is approved for this service. All recessed incandescent fixtures must have thermal protection. This restriction is not required if the fixture is installed in poured concrete and is approved for this use or when the fixture is listed and by construction design provides thermally protected characteristics.

410-66 Clearance and Installation

Recessed parts of the fixture must be at least ½ in. from combustible material and at least 3 in. from thermal insulation unless the fixture is approved for direct contact with the insulation.

410-67 Wiring

The insulation used on conductors must be suitable for the temperatures that they will be subjected to. Tap conductors can run to a box at least 1 ft. from the fixture. The raceway used for the tap conductor must be at least 4 ft. long and not more than 6 ft. long, be a suitable raceway or the tap conductor must be Type AC or MC cable.

N. Construction of Flush and Recessed Fixtures

410-68 Temperature

410-69 Housing

410-70 Lamp Wattage Marking

410-71 Solder Prohibited

410-72 Lampholders

P. Special Provisions for Electric-Discharge Lighting Systems of 1000 Volts or Less

410-73 General

410-74 Direct-Current Equipment

410-75 Open-Circuit Voltage Exceeding 300 Volts

410-76 Fixture Mounting

410-77 Equipment Not Integral with Fixture

410-78 Autotransformers

410-79 Switches

Q. Special Provisions for Electric-Discharge Lighting Systems of More Than 1000 Volts

R. Lighting Track

ARTICLE 411

LIGHTING SYSTEMS OPERATING AT 30 VOLTS OR LESS

411-3 Listing Required

411-4 Locations Not Permitted

411-5 Secondary Circuits

411-6 Branch Circuit

411-7 Hazardous (Classified) Locations

ARTICLE 422
APPLIANCES

A. General

422-1 Scope

This covers electric appliances used in any type of occupancy.

422-2 Live Parts

No live parts can be normally exposed. An exception to this is in toasters, grills, and the like when current-carrying parts must be exposed at high temperatures.

422-3 Other Articles

All other parts of the Code apply when applicable. When used in hazardous areas, Articles 500 through 517 apply. Article 430 applies to motor appliances and Article 440 applies to appliances with hermetic refrigerant motors.

B. Branch-Circuit Requirements

422-4 Branch-Circuit Rating

Generally, the rating of the circuit must be at least that of the total rating of the appliance for an individual appliance branch circuit. There are three exceptions. Motor circuits are rated higher in accordance with Part B of Article 430. If the appliance is to be continuously loaded (not a motor), the branch circuit must be rated at least 125% of the appliance unless the branch-circuit device and assembly are rated for continuous loading at 100%, in which case the branch circuit cannot be less than 100% of the appliance. Branch circuits for household cooking appliances are to be rated as provided in Table 220-19. If the branch circuit supplies an appliance and other loads, use the provisions of Section 210-23.

422-5 Branch-Circuit Overcurrent Protection

Must be protected in accordance with Section 240-3. If a protective device rating is marked on an appliance, this rating must not be exceeded.

C. Installation of Appliances

422-6 General

422-7 Central Heating Equipment

Must be connected to an individual branch circuit except for fixed electric space heating equipment.

422-8 Flexible Cords

Specific types of cords are indicated for specific uses, with noted exceptions.

422-9 Cord- and Plug-Connected Immersion Heaters

422-10 Protection of Combustible Materials

422-11 Stands for Cord- and Plug-Connected Appliances

422-12 Signals for Heated Appliances

422-13 Flatirons

422-14 Water Heaters

422-15 Infrared Lamp Industrial Heating Appliances

422-16 Grounding

422-17 Wall-Mounted Ovens and Counter-Mounted Cooking Units

These appliances can be installed with a plug and receptacle combination only for ease in servicing or installation. If a plug and receptacle is used, it is not to be considered the disconnecting means required and they must be approved for the temperature in the space.

422-18 Support of Ceiling Fans

422-19 Other Installation Methods

D. Control and Protection of Appliances

422-20 Disconnecting Means

This must be provided as described in this part. If supplied by more than one source, all disconnects must be grouped.

422-21 Disconnection of Permanently Connected Appliances

If the device is not over 300 V A or ⅛ hp, the branch-circuit overcurrent device can be the disconnecting means. If the ratings are higher, the branch-circuit overcurrent device can only be used as the disconnecting means if it can be locked in the open position and is in sight of the user. Refer to Section 422-27 for more than ⅛-hp motor-driven appliances and Section 422-25 for unit-switched appliances.

422-22 Disconnection of Cord- and Plug-Connected Appliances

For these types of appliances an accessible separable connector or plug and receptacle can be used as the disconnecting means. Where it is not accessible a disconnecting means in accordance with Section 422-21 must be used. In the case of the household range, the plug and receptacle must be accessible from the front either directly or through removal of a drawer. The receptacle or connector must be rated at least the same as the appliances. Demand factors are permitted if authorized in other sections of the *NEC*®. The attachment plugs and connectors must be constructed so that live parts cannot be inadvertently touched; and designed so as to be able to interrupt the rated current and not to fit into a receptacle of lesser rating.

422-23 Polarity in Cord- and Plug-Connected Appliances

The attachment plug must be polarized or of the grounding type if the appliance has a single-pole line connected manually operated switch; an Edison-base lampholder; or a 15- or 20-A receptacle. Polarity of Edison-base lampholders are noted in Section 410-42(a). There is an exception for a hand-held double insulated shaver.

422-24 Cord- and Plug-Connected Appliances Subject to Immersion

A portable free-standing hydromassage unit or hand-held hair dryer has to be made to protect the user from electrocution if the unit becomes immersed when in the on or off position.

422-25 Unit Switch(es) as Disconnecting Means

A unit switch or switches that is part of the appliance can be considered the disconnecting means for the appliance if it has a position marked ``off,'' disconnects all ungrounded conductors and other disconnecting means are installed as follows:

(a) In multifamily dwellings this other disconnecting means must be installed at least on the same floor that the appliance is installed. It can control other appliances and lamps.

(b) In two-family dwellings this other disconnecting means can be installed outside the dwelling unit and it can be a switch for the dwelling unit, such as in a panel in the basement.

(c) The service disconnect will fulfill this other requirement in a one-family dwelling.

(d) In all other occupancies, if readily accessible, the branch unit switch or circuit breaker can be used as the other disconnecting means.

422-26 Switch and Circuit Breaker to Be Indicating

422-27 Disconnecting Means for Motor-Driven Appliances

In the case of a permanently installed motor-driven appliance that is larger than ⅛ hp, a switch or circuit breaker can be used if it is in sight of the appliance. It does not have to be in sight if it is the other disconnecting means, as noted in Section 422-25(a), (b), or (c).

422-28 Overcurrent Protection

(a) Appliances. They are permitted to be protected when they have branch circuits as indicated in Sections 422-5 and 422-4 and paragraphs (b) and (f) in this section. Motors are excepted and must comply with Articles 430 and 440 where applicable.

(b) Household-Type Appliance with Surface Heating Elements. If its demand is more than 60 A when calculated with Table 220-19, it must have its power supply subdivided with no circuit more than 50 A.

(c) Infrared Lamp Commercial and Industrial Heating Appliances. Cannot be greater than 50 A.

(d) Open-Coil or Exposed Sheathed-Coil Types of Surface Heating Elements in Commercial-Type Heating Appliances. Cannot be greater than 50 A.

(e) Single Nonmotor-Operated Appliance. The branch-circuit overcurrent device rating should not exceed the protective device rating noted on the unit. If no rating is noted, use 20 A or less if the unit rating is not more than 13.3 A and 150% or less of the unit rating if it is more than 13.3 A.

(f) Electric Heating Appliances Employing Resistance-Type Heating Elements Rated More Than 48 Amperes. Factory-installed overcurrent devices which must be used in this case are discussed in this paragraph.

E. Marking of Appliances

422-29 Cord- and Plug-Connected Pipe Heating Assemblies

If it is intended to prevent freezing of piping must be listed.

422-30 Nameplate

An easily accessible or visible nameplate with a name and rating in volts and amperes or volts and watts must be installed. Frequency must be noted when applicable.

422-31 Marking of Heating Elements

They must be marked if rated over 1 A, are replaceable in the field, and are part of an appliance.

422-32 Appliances Consisting of Motors and Other Loads

ARTICLE 424
FIXED ELECTRIC SPACE HEATING EQUIPMENT

A. General

424-1 Scope

This article does not apply to heating and room air conditioning. It does apply to heating cables, unit boilers, central systems, and other fixed electric space heating equipment.

424-2 Other Articles

All applicable rules of the *NEC*® apply.

424-3 Branch Circuits

(a) Branch-Circuit Requirements. It can be any size as long as it is an individual branch circuit supplying only one appliance. If more than one appliance is on the circuit, 30 A is the largest rating permitted (50 A for nonresidential infrared heating equipment).

(b) Branch-Circuit Sizing. Generally, the branch-circuit size and overcurrent rating must be at least 125% of the rating of the appliance. There are a few exceptions. These include contactors and relays [refer to Section 210-22(c), Exception No. 2], and fixed electric space heating equipment, including hermetic refrigeration equipment (refer to Sections 440-34 and 440-35).

B. Installation

424-9 General

424-10 Special Permission

424-11 Supply Conductors

424-12 Locations

This equipment cannot be installed where it may be severely damaged unless protected. When installed in damp or wet locations, the equipment has to be approved for such locations.

424-13 Spacing from Combustible Materials

Adequate spacing must be provided unless approved for direct contact.

424-14 Grounding

Exposed metal parts which will probably become energized must be grounded in accordance with Article 250.

C. Control and Protection of Fixed Electric Space Heating Equipment

424-19 Disconnecting Means

This must be provided. When more than one source is used, all disconnecting means must be grouped and marked.

(a) Heating Equipment with Supplementary Overcurrent Protection. Must be in sight of the supplementary overcurrent device and on the supply side of the overcurrent device if fuses and in addition item 1 or 2 following.

1. *Heater containing no motor rated over ⅛ horsepower.* This disconnecting means or a unit switch complying with Section 424-19(c) can be used as the disconnecting means for the motor controller and heater if it is in sight of the motor controller and heater or can be locked in the open position.

2. *Heater containing a motor(s) rated over ⅛ horsepower.* The same holds true as noted in the preceding subparagraph (1). In addition, if it is not in sight of the heater, a separate disconnect complying with Section 424-19(c) can be installed. If it is not in sight of the motor controller location, a disconnect means must be installed in accordance with Section 430-102, and where the motor is not in sight of the controller must comply with Section 430-102(b).

(b) Heating Equipment without Supplementary Overcurrent Protection

1. *Without motor or with motor not over ⅛ horsepower.* The branch-circuit switch or circuit breaker can serve as the disconnecting means if the switch or circuit breaker can be locked in the open position or is within sight of the heater.

2. *Over ⅛ horsepower.* A disconnecting means must be located within sight of the motor controller except when installed in accordance with Section 424-19(a)(2).

(c) Unit Switches as Disconnecting Means. A unit switch that is part of the heater disconnects all ungrounded conductors and has a marked off position can be used as the disconnecting means if other disconnecting means are installed as follows:

1. In multifamily family dwellings, this disconnecting means must be installed at least on the same floor as the heater. It can control other heaters and lamps.
2. In two-family dwellings this disconnecting means can be installed either inside or outside the dwelling unit. There is also a provision for additional control.
3. In one-family dwellings the service disconnect can serve this purpose.
4. In other occupancies, if readily accessible, the branch-circuit switch or circuit breaker can be used.

424-20 Thermostatically Controlled Switching Devices

These devices can also serve as the disconnecting means if all the following conditions are met; have a marked off position, interrupt all ungrounded conductors when manually put in that position, cannot be automatically energized in the off position, and are located in accordance with Section 424-19.

424-21 Switch and Circuit Breaker to Be Indicating

424-22 Overcurrent Protection

Branch-circuit protection as provided in Article 210 can be used. Motors must comply with Articles 430 and 440. Resistance element cannot be protected at a higher rating than 60 A. Equipment rated at more than 48 A must be subdivided, with each subdivision rated not more than 48 A. The manufacturers must provide the division and overcurrent protection for the division.

Field conductors between the heater and supplementary overcurrent protection must be sized at 125% of the load if the heater is rated less than 50 kW.

If the heater is rated at least 50 kW the conductors can be sized at 100% of the rating if the heater is marked for 100% conductor rating and the conductors are not smaller than the marked rating. Also, a temperature control device must be installed.

D. Marking of Heating Equipment

424-28 Namplate

Must have proper marking information: volts, amperes, horsepower, etc.

424-29 Marking of Heating Elements

Volts and amperes or volts and watts.

E. Electric Space Heating Cables

424-34 Heating Cable Construction

Nonheating cables of at least 7 ft. must be installed at the factory.

424-35 Marking of Heating Cables

Proper marking is required. Special color coding is required for different voltages of heating cables: 120 V, yellow; 208 V, blue; 240 V, red; 277 V. brown.

424-36 Clearances of Wiring in Ceilings

Two inches of space is required above a heated ceiling. An operating ambient temperature of 50°C is to be used for all sizing and appropriate correction factors used. If located above thermal insulation 2 in. in thickness, the temperature correction is not needed.

424-37 Location of Branch-Circuit and Feeder Wiring in Walls

Comply with Article 300 and Section 310-10.

424-38 Area Restrictions

The cables cannot go beyond the room in which they originate. They cannot be installed in closets or over a partition that extends to be ceiling. There is also a restriction as to how they may be run over cabinets. They may be used in closet ceilings for low-temperature sources for humidity control, if there are no obstructions between floor and ceiling in that part of the ceiling.

424-39 Clearance from Other Objects and Openings

The heating cables must be at least 8 in. from boxes used for surface-mounted lighted fixtures and 2 in. from recessed lighting fixtures. They cannot be installed where they are covered by any surface mounted unit.

424-40 Splices

The cable's length cannot be changed.

424-41 Installation of Heating Cables on Dry Board, in Plaster and on Concrete Ceilings

(a) Cables are not permitted in walls unless it is a single run to reach a dropped ceiling.

(b) A minimum spacing of 1½ in. is required for cables rated at 2¾ W per foot and under for adjacent runs.

(c) Heating cables cannot be applied directly to metal surfaces. They can be only to fire-resistant surfaces. A coat of plaster must be applied to a metal surface.

(d) The entire heating cable, at least 3 in. of the nonheating cable and the splice between them must be embedded in plaster or dry board.

(e) The surface of the ceiling must have a coating of at least ½ in. of plaster or other approved material.

(f) Cables have to be secured at a maximum of 16 in. on center unless approved for 6 in. on center.

(g) When dry board is used for the ceiling, it cannot be thicker than ½ in.

(h) Cables cannot come in contact with metal or other surfaces which can conduct electricity.

(i) The cables must be installed parallel to joists in a dry board installation and at least 2½ in. between center of adjacent runs of cable under the center line of the joist. Care must be taken not to drive nails through the heating cables.

424-42 Finished Ceilings

Coverings cannot have thermal insulating properties. Paint and wallpaper can be used.

424-43 Installation of Nonheating Leads of Cables and Panels

424-44 Installation of Cables in Concrete or Poured Masonry Floors

424-45 Inspection and Tests

F. Duct Heaters

424-57 General

424-58 Identification

424-59 Airflow

424-60 Elevated Inlet Temperature

Duct heaters must be approved for mounting in ducts and at the expected temperatures, as well as assure proper airflow.

424-61 Installation of Duct Heaters with Heat Pumps and Air Conditioning

424-62 Condensation

424-63 Fan Circuit Interlock

424-64 Limit Controls

424-65 Location of Disconnecting Means

Must be within sight of the controller except as noted in Section 424-19(a).

424-66 Installation

G. Resistance-Type Boilers

424-70 Scope

424-71 Identification

424-72 Overcurrent Protection

424-73 Over-Temperature Limit Control

424-74 Over-Pressure Limit Control

424-75 Grounding

H. Electrode-Type Boilers

424-80 Scope

424-81 Identification

424-82 Branch-Circuit Requirements

424-83 Over-Temperature Limit Control

424-84 Over-Pressure Limit Control

424-85 Grounding

424-86 Markings

B. Electric Radiant Heating Panels and Heating Panel Sets

424-90 Scope

424-91 Definition

424-92 Markings

424-93 Installation

424-94 Clearances of Wiring in Ceilings

424-95 Location of Branch-Circuit and Feeder Wiring in Walls

424-96 Connection to Branch-Circuit Conductors

424-97 Nonheating Leads

424-98 Installation in Concrete or Poured Masonry

424-99 Installation under Floor Covering

ARTICLE 426
FIXED OUTDOOR ELECTRIC DE-ICING AND SNOW-MELTING EQUIPMENT

A. General

426-1 Scope

This article covers installations both embedded and exposed.

426-2 Definitions

426-3 Application of Other Articles

The rest of the *NEC*® is applicable unless changed by this article.

426-4 Branch-Circuit Sizing

Conductors and overcurrent protection must be at least 125% of the load. Section 240-3 rating or setting is permitted.

B. Installation

426-10 General

426-11 Use

426-12 Thermal Protection

Must be protected against contact by persons if operated at temperatures above 140°F.

426-13 Identification

Warning signs must be posted where used.

426-14 Special Permission

C. Resistance Heating Elements

426-20 Embedded De-icing and Snow-Melting Equipment

Units cannot exceed 120 W per square foot and must have at least 1 in. between cables. They can be installed either on an asphalt or masonry base at least 2 in. thick and covered with at least 1½ in. of asphalt or masonry, or on any approved type of base embedded in 3½ in. of asphalt or masonry and covered with at least 1½ in. of asphalt or masonry. They must be secured and not bridge expansion joints unless proper provisions are made.

426-21 Exposed De-icing and Snow-Melting Equipment

They must be secured, proper temperature limiting, and have no cross-bridge expansion joints unless provided for.

426-22 Installation of Nonheating Leads for Embedded Equipment

426-23 Installation of Nonheating Leads for Exposed Equipment

426-24 Electrical Connection

426-25 Marking

426-26 Corrosion Protection

426-27 Grounding

426-28 Equipment Protection

D. Impedance Heating

426-30 Personnel Protection

426-31 Voltage Limitations

The secondary winding of the isolation transformer must not operate at greater than 30 V ac. If ground fault circuit interrupter is used—80 V ac.

426-32 Isolation Transformer

426-33 Induced Currents

426-34 Grounding

E. Skin Effect Heating

426-40 Conductor Ampacity

426-41 Pull Boxes

426-42 Single Conductor in Enclosure

426-43 Corrosion Protection

426-44 Grounding

F. Control and Protection

426-50 Disconnecting Means

Must have a disconnecting means that opens the ungrounded wires. The branch-circuit breaker can be used if readily accessible. Also permitted is a factory installed attachment plug if it is rated not more than 20 A and 150 V to ground.

426-51 Controllers

If the temperature controller operates line current and has an off position, it must open all ungrounded conductors. It cannot be used as a disconnecting means unless it can be locked in the open position. If no off position is provided, it does not have to open all ungrounded conductors and cannot be used as a disconnecting means. If it controls a relay, it does not have to open all ungrounded conductors and cannot be used as a disconnecting means. There are also rules in the *NEC*® for combined switching devices.

426-52 Overcurrent Protection

Refer to Section 426-4.

426-54 Cord- and Plug-Connected De-icing and Snow Melting Equipment.

ARTICLE 427
FIXED ELECTRIC HEATING EQUIPMENT FOR PIPELINES AND VESSELS

A. General

427-1 Scope

427-2 Definitions

427-3 Application of Other Articles

427-4 Branch-Circuit Sizing

B. Installation

427-10 General

427-11 Use

427-12 Thermal Protection

427-13 Identification

C. Resistance Heating Elements

427-14 Secured

427-15 Not in Direct Contact

427-16 Expansion and Contraction

427-17 Flexural Capability

427-18 Power Supply Leads

427-19 Electrical Connections

427-20 Marking

427-21 Grounding

427-22 Equipment Protection

427-23 Metal Covering

D. Impedance Heating

427-25 Personnel Protection

427-26 Voltage Limitations

427-27 Isolation Transformer

427-28 Induced Currents

427-29 Grounding

427-30 Secondary Conductor Sizing

E. Induction Heating

427-35 Scope

427-36 Personnel Protection

427-37 Induced Current

F. Skin Effect Heating

427-45 Conductor Ampacity

427-46 Pull Boxes

427-47 Single Conductor Enclosure

427-48 Grounding

G. Control and Protection

427-55 Disconnecting Means

427-56 Controls

427-57 Overcurrent Protection

ARTICLE 430 MOTORS, MOTOR CIRCUITS, AND CONTROLLERS

A. General

430-1 Scope

This article covers requirements for motors, motor branch circuit and feeder conductors and their protection, motor overload protection, motor control circuits, motor controllers, and motor control centers. There are general and specific requirements. Diagram 430-1 in the *NEC*® describes where in Article 430 different parts of the circuits are covered as well as other articles. This article does not cover air-conditioning and refrigeration equipment, which is covered in Article 440 and motor control centers which is covered in Section 384-4.

430-2 Adjustable-Speed Drive Systems

430-3 Part Winding Motors

430-5 Other Articles

There are numerous other articles with which motor and controllers must comply for special cases.

430-6 Ampacity and Motor Rating Determination

Generally, Tables 310-16 through 310-19 or calculations per Section 310-15(b) are to be used for conductor selection and Tables 430-147 through 430-150 are used to determine ampere rating for sizes of wire switches, circuit protection, and the like (not the nameplate ratings). They are not used for separate motor overload protection. For this, the nameplate ratings are used. Exceptions to this rule are multispeed motors, shaded pole motors, permanent-split-capacitor-type fan or blower motors, torque motors, and ac adjustable voltage motors.

430-7 Marking on Motors and Multimotor Equipment

430-8 Marking on Controllers

430-9 Terminals

430-10 Wiring Space in Enclosures

Table 430-10(b) indicates the minimum wire bending space at the terminals of enclosed motors.

430-11 Protection against Liquids

430-12 Motor Terminal Housings

Tables 430-12(b), 430-12(c)(1), and 430-12(c)(2) give various dimension and space requirements. There is also information on connections.

430-13 Bushing

430-14 Location of Motors

430-16 Exposure to Dust Accumulations

430-17 Highest Rated or Smallest Rated Motor

If this is to be determined for compliance with Sections 430-24, 430-53(b), and 430-53(c), use the rated full-load current from Tables 430-147, 430-148, 430-149, and 430-150.

430-18 Nominal Voltage of Rectifier Systems

B. Motor Circuit Conductors

430-21 General

The rules of this section apply to circuits supplying motors at 600 V or less. Section 430-124 applies to motors over 600 V.

430-22 Single Motor

Generally size conductors at 125% of the motor full-load rated current for the supply to a single motor. There are two exceptions to this rule. The first is for short-time intermittent, periodic, or varying duty. For these, use the values in Table 430-22(a). The second is for direct-current motors that operate from a single-phase power supply which is rectified. There are regulations for multispeed motors and wye start and delta run motors. When the junction box is allowed to be separated from the motor, a special exception is made for wire size on motors of 1 hp or less.

430-23 Wound-Rotor Secondary

The secondary leads for a continuous-duty wound-rotor motor must be rated at least 125% of the full-load secondary current. If the motor is not continuous, Table 430-22(a) must be used. Table 430-23(c) is used when the secondary resistor is separate from the controller.

430-24 Several Motors or a Motor(s) and Other Load(s)

In this case the conductors are rated at 125% of the full load current of the largest motor (the same as a single motor) plus the full-load current of all additional motors plus the ampere rating of the other loads which are calculated in accordance with Article 220 and other sections of the *NEC®*. There are three exceptions, continuous-duty, motor-operated fixed space heating, and interlocked motors.

430-25 Multimotor and Combination-Load Equipment

The conductors for these loads are based, as a minimum, on the circuit ampacity marked on the equipment in accordance with Section 430-7(d). If the equipment is not factory wired, and the nameplates visible as per Section 430-7(d)(2), the conductors are calculated in accordance with Section 430-24.

430-26 Feeder Demand Factor

If all motors will not operate simultaneously for any number of reasons, the authority having jurisdiction can allow the use of a demand factor.

430-27 Capacitors with Motors

430-28 Feeder Taps

Feeder tap conductors must be sized at least according to Part B of this section and either be enclosed and 10 ft. or less in length or be one-third or more of the rating of the feeder, 25 ft. or less in length, and protected. There is an exception for feeder taps over 25 ft. in high bay manufacturing buildings. There are additional rules.

430-29 Constant-Voltage DC Motors—Power Resistors

C. Motor and Branch-Circuit Overload Protection

430-31 General

The overload protection is usually the thermal overloads in the controller or starter. It should not be confused with the short-circuit protection, which is the fuse or circuit breaker in the panelboard. Diagram 430-1 in the Code should be referred to, as well as Example 8, Chapter 9.

430-32 Continuous-Duty Motors

There are four methods acceptable:

(a) More Than 1 Horsepower

1. A separate overload device must be used. It must be rated at not more than 125% of the motor nameplate rating if the motor is rated with a temperature rise of not more than 40°C or a service factor of 1.15 or more. The rating is reduced to 115% for all other motors. Exceptions are provided in Section 430-34.
2. Thermal protectors as part of the motor are listed in this paragraph.
3. If the motor is part of an approved assembly, will not normally have overloads, and has some protection against failure to start, additional overload protection is not necessary.
4. Motors larger than 1500 hp are discussed in this paragraph.

(b) One Horsepower or Less, Nonautomatically Started. If it is within sight of the controller location, the ground-fault protection and branch-circuit short-circuit protection is sufficient. The sizings noted in Part D are to be used. It can be on a 120-V circuit rated 20 A or less. If not within sight of the controller or permanently installed, it must comply with Section 430-32(c), which follows.

(c) One Horsepower or Less, Automatically Started. Here again there are four acceptable methods; the first three are similar to the first three for motors rated more than 1 hp. These include separate devices (ratings are indicated), thermal protectors, and cases where no additional protection other than branch-circuit protection is required. A fourth case is that where the impedance of the motor protects against overheating when the motor does not start.

(d) Wound-Rotor Secondaries. These can have a motor overload device.

430-33 Intermittent and Similar Duty

You can omit the separate overload protection for these types of motors and use only the branch-circuit, short-circuit and ground-fault protective device.

430-34 Selection of Overload Relay

In the event the ratings chosen as listed in Section 430-32(a)(1) and (c)(1) will not allow the motor to start, this section gives larger permissible ratings.

430-35 Shunting during Starting Period

There are cases where the overload protection can be shunted during starting.

430-36 Fuses—In Which Conductor

When used as overload protection, put one in each ungrounded conductor and also the grounded conductor if the supply is three-phase, three-wire ac with one phase grounded.

430-37 Devices Other Than Fuses—In Which Conductor

Table 430-37 gives the requirements.

430-38 Number of Conductors Opened by Overload Devices

The number of conductors to be opened must prevent current to the motor for devices other than fuses or thermal protection.

430-39 Motor Controller as Overload Protection

This is permitted if it is in accordance with Table 430-37 and operator in the running position and also in the starting position for an ac motor.

430-40 Overload Relays

If the device used for motor overload protection cannot clear short circuits, they must be protected by either fuses or circuit breakers or a motor short-circuit protector sized as noted in Section 430-52. There are two exceptions to the sizing rule—group installations and where an overload relay is used—both where approved and the rating is marked in the equipment.

430-42 Motors on General-Purpose Branch Circuits

This section indicates what overload protection must be used when motors are connected to general-purpose branch circuits. It must also comply with Article 210.

(a) Not Over 1 Horsepower. Can be connected without overload protection if the rules of Section 430-32(b) and (c) and Section 430-53(a)(1) and (a)(2) are followed.

(b) Over 1 Horsepower. Each motor must have overload protection as noted in Section 430-32.

(c) Cord- and Plug-Connected. If no overload protection is provided, as in (a) above, the ratings of the plug and receptacle cannot be larger than 15 A 125 V or 10 A 250 V. If protection is provided as in (b) above, the overload device must be part of the motor or appliance.

(d) Time Delay. Time delay must be provided in the branch-circuit and ground-fault devices to let the motor start.

430-43 Automatic Restarting

430-44 Orderly Shutdown

If an orderly shutdown is necessary to protect people rather than a quick opening of the circuit, the overload device can be connected to an alarm rather than open the circuit under the provisions of this section.

D. Motor Branch-Circuit, Short-Circuit, and Ground-Fault Protection

430-51 General

430-52 Rating or Setting for Individual Motor Circuit

The device must carry the starting current. Table 430-152 lists the maximum settings of the protective devices. If the values in the tables are not a standard-size device, the rating can be increased to the next highest rating. If the rating in the table will not permit the motor to start, the rating can be increased as follows:

1. Nontime delay fuse less than 601 A up to 400% of full-load current
2. Time delay fuses up to 225% of full-load current
3. Circuit breakers (inverse time) up to 400% for full-load currents of up to 100 A and 300% for larger currents
4. Fuses between 601 and 6000 A up to 300% of full-load current

Also listed in this section are the rules for when instantaneous trip circuit breakers and motor short-circuit protectors can be used; rules for multispeed motors; rules for torque motors; and rules for phase converters. Where a maximum rating is indicated by the manufacturer, it cannot be exceeded. The reader is referred to this section of the Code for complete information.

430-53 Several Motors or Loads on One Branch Circuit

This section indicates under what conditions two or more motors or combined loads are permitted on one branch circuit.

(a) If none of the motors is greater than 1 hp, the circuit is protected at 20 A at 120 V or not more than 15 A at 600 V, the full-load current of each motor is not more than 6 A, individual overload protection meets the requirements of Section 430-32, and the rating marked on any controller for branch-circuit short circuit and ground-fault protection is not exceeded, more than one motor or combination loads are permitted on one branch circuit.

(b) When the branch-circuit, short-circuit and ground-fault protective devices rating is not larger than required for the smallest motor and all motors have individual overload protection and the combination load will operate under these conditions, more than one motor or combination loads are permitted.

(c) Group installations of more than one motor or combination loads are permitted if the overload devices and motor controllers are in a listed factory assembly, and motor short circuit and ground-fault protection are in the assembly or noted on the assembly; or if the controllers, overload devices, branch circuit, short circuit, and ground-fault devices are field-installed separately with manufacturer's instructions for use with each other and the following conditions are met:

1. The overload device must be approved for group installation and have a specified maximum fuse or inverse time circuit breaker rating, or both.
2. The motor controller for each motor must comply with the same rules as in (1) above.
3. The circuit breakers must be approved for group installation and have inverse time characteristics.
4. The rating of the fuse or circuit breaker for the branch

circuit must comply with rules of Sections 430-52 and 240-3. Exception 1. It must be at least that of Section 430-52 for the largest motor and the sum of full-load ratings of other motors and loads.

5. The rating allowed in Section 430-40 for the smallest motor will also be the maximum for the branch-circuit fuses or inverse time circuit breakers.

(d) In a group installation as described in (c) above, if the conductors to any single motor are not larger than the branch-circuit conductors or are not less than one-third the rating of the branch-circuit conductors in accordance with Section 430-22, and not farther than 25 ft. from the controller and physically protected, then installation will not require individual branch-circuit short-circuit and ground-fault protection.

430-54 Multimotor and Combination-Load Equipment

The branch-circuit short-circuit and ground-fault device rating must not be larger than the rating marked on the equipment and comply with Section 430-7(d).

430-55 Combined Overcurrent Protection

As long as the rating is not larger than that required for overload protection in Section 430-32, the branch-circuit short-circuit and ground-fault protection and overload protection can be in one device.

430-56 Branch-Circuit Protective Devices—In Which Conductor

430-57 Size of Fuseholder

430-58 Rating of Circuit Breaker

E. Motor Feeder Short-Circuit and Ground-Fault Protection

430-61 General

430-62 Rating or Setting—Motor Load

The rating must be that of the largest branch-circuit protective device plus the full-load currents of all other motors. If larger feeders have been installed for future expansion, the rating can be that of the feeder. Reference is made to Sections 430-24 and 440-22(a) and Table 430-152.

430-63 Rating or Setting—Power and Light Loads

When combined loads are installed in one feeder, the rating can be the combination of that permitted by Articles 210 and 220 for lighting and appliances plus that permitted by Section 430-52 for a single motor and Section 430-62 for more than one motor.

F. Motor Control Circuits

430-71 General

The motor control circuit carries the electric signal to control the controller but not power for the circuit.

430-72 Overcurrent Protection

If the circuit is tapped from the load side of a branch-circuit protective device, it must have protection as specified in Section 430-72. If not, it must be protected as specified in Section 725-23. Values for overcurrent protection for conductors are specified in Table 430-72(b). There are four exceptions to this rule.

There are also requirements for a control circuit transformer.

430-73 Mechanical Protection of Conductor

430-74 Disconnection

Motor control circuits must have a disconnecting means. One device can disconnect both the motor and control circuits from power, or separate ones can be used. If a transformer is used to reduce the voltage for the control circuit and it is located in the controller enclosure, the disconnecting means must be in the supply side of the transformer. There are a number of exceptions.

G. Motor Controllers

430-81 General

The word "controller" is defined in Article 100. Here it can mean a device or switch to stop and start a motor by breaking the motor circuit current. When the motor is stationary and of not more than ⅛ hp, the branch-circuit protective device can be used if the motor is normally left running and cannot be damaged by not starting or by overloads. The plug and receptacle can be the controller if the motor is portable and not more than ⅓ hp.

430-82 Controller Design

430-83 Ratings

Generally, it must have a horsepower rating at least equal to that of the motor. There are five exceptions.

430-84 Need Not Open All Conductors

Unless it serves as a disconnecting device, it does not have to open all conductors.

430-85 In Grounded Conductors

This is permitted if it also simultaneously opens all other conductors.

430-87 Number of Motors Served by Each Controller

Each motor must normally have its own controller. There are exceptions to this rule if the motors are rated at not more than 600 V and the single controller is rated at not less than the sum of all motors it will control. Under these conditions it is allowed if more than one motor is used in one device or one overcurrent device is used for a group of motors [Section 430-53(a)] or the motors are in one room in sight of the controller.

430-88 Adjustable-Speed Motors

430-89 Speed Limitation

430-90 Combination Fuseholder and Switch as Controller

430-91 Motor Controller Enclosure Types

H. Motor Control Centers

430-92 General

A motor control center is an assembly of enclosed sections which principally contain motor control units and have a common power bus.

430-94 Overcurrent Protection

They must have overcurrent protection, either an overcurrent device upstream or a main device in the motor control center. It must be in accordance with Article 240 and sized in accordance with the common power bus.

430-95 Service-Entrance Equipment

430-96 Grounding

All sections must be bonded with an equipment grounding conductor or grounding bus in accordance with Table 250-95. All equipment grounding conductors must terminate here. In the case of a single section a grounding termination point must be provided.

430-97 Busbars and Conductors

(a) Support and Arrangement. They must be protected from damage and held firmly. Only the conductors connected or used for control are allowed in a section.

(b) Phase Arrangement. Three phase buses are arranged A, B, left to right, top to bottom, or front to back when standing in the front.

(c) Minimum Wire Bending Space. As noted in Article 373.

(d) Spacings. As noted in Table 430-97.

(e) Barriers. Used in service-entrance motor control centers.

430-98 Marking

I. Disconnecting Means

430-101 General

430-102 Location

It must be in sight from the controller location. An exception is made for motors over 600 V if it can be locked out and a warning sign is posted. Also, one disconnect can be used for a group of controllers for a multimotor device. A disconnecting means must be located within sight of the motor location except in accordance with Section 430-102(a).

430-103 Operation

430-104 To Be Indicating

430-105 Grounded Conductors

If all conductors are opened simultaneously, one pole can be in a grounded conductor.

430-107 Readily Accessible

430-108 Every Switch

430-109 Type

Generally, it must be a horsepower-rated motor circuit switch, a circuit breaker, or a molded-case switch. There are seven exceptions.

Exception No. 1. A motor circuit switch for a Design E motor 2 horsepower or more must be marked for a Design E motor and it must have a horsepower rating 1.4 times or more than the rating of a 3 through 100 horsepower motor or 1.3 times or more the rating of a motor larger than 100 horsepower.

Exception No. 2. The branch-circuit overcurrent device can be used for stationary motors of ⅛ hp or less.

Exception No. 3. Stationary motors 2 hp or less and 300 V or less can have a general-use switch rated twice that of the motor. A general-use snap switch (ac only) can be used on ac circuits if the motor is not more than 80% of the rating of the switch.

Exception No. 4. Motors over 2 hp up to 100 hp. A general-use switch can be used for autotransformer controllers if the motor drives a generator with overload protection; the controller can interrupt locked rotor current, has no voltage release, and has running overload protection not larger than 125% of full-load current; separate fuses or inverse time circuit breakers are provided and rated not more than 150% of full-load current.

Exception No. 5. Stationary motors more than 40 hp dc or 100 hp ac can have a general use or isolating switch if marked "Do Not Operate Under Load." The code has requirements for Design E motors.

Exception No. 6. Cord- and plug-connected motors can use

horsepower-rated attachment plugs and receptacles rated at least that of the motor. This is not required for cord and plug appliances (Section 422-22), room air conditioners (Section 440-63), or a portable motor ⅓ hp or less.

Exception No. 7. A general-use switch can be used for a torque motor.

Exception No. 8. An instantaneous trip circuit breaker which is part of a listed combination motor controller can serve as a disconnect.

430-110 Ampere Rating and Interrupting Capacity

(a) It must have a rating of at least 115% of the full-load current rating of the motor.

(b) It must have a rating of at least 115% of the nameplate current for torque motors.

(c) For combination loads of motors or motors and other types of loads, the ratings of the disconnect must not be less than 115% of the total of the full-load current of all the loads. The code makes reference to the method of determining the combined load.

430-111 Switch or Circuit Breaker as Both Controller and Disconnecting Means

A switch or circuit breaker can be used as the disconnect and controller if it complies with Section 430-83: if it opens all ungrounded conductors to the motor, has an overcurrent protective device which opens all ungrounded conductors to the switch, and is either an air-break switch an inverse time circuit breaker, or an oil switch.

430-112 Motors Served by Single Disconnecting Means

Generally, a motor must have its own disconnecting means. There is an exception.

430-113 Energy from More Than One Source

Generally, each source must have a disconnect. There are two exceptions.

J. Over 600 Volts, Nominal

430-121 General

430-122 Marking on Controllers

430-123 Conductor Enclosures Adjacent to Motors

430-124 Size of Conductors

430-125 Motor Circuit Overcurrent Protection

430-126 Rating of Motor Control Apparatus

430-127 Disconnecting Means

K. Protection of Energized Parts—All Voltages

430-131 General

430-132 Where Required

If operating at 50 V or more, the exposed live parts must be guarded against accidental contact by enclosure or location. This can be done in a special room or enclosure, on a balcony or by installation at least 8 ft. above the floor.

430-133 Guards for Attendants

If the only guards are by location, as noted in Section 430-132, then insulating mats or platforms are required.

L. Grounding All Voltages

430-141 General

430-142 Stationary Motors

The frames must be grounded if in a wet location and not isolated, supplied by wire in a metal enclosure, in a hazardous location, or if motor operates over 150 V to ground.

430-143 Portable Motors

430-144 Controllers

430-145 Method of Grounding

ARTICLE 440
AIR CONDITIONING AND REFRIGERATING EQUIPMENT

A. General

440-1 Scope

This section states the scope.

440-2 Definitions

440-3 Other Articles

Many additional articles are listed, such as 422, 424, 430, 460, and 470, and some not in Chapter 4.

440-4 Marking on Hermetic Refrigerant Motor-Compressors and Equipment

440-5 Marking on Controllers

440-6 Ampacity and Rating

The rating on the nameplate is to be used for a hermetic refrigerant motor compressor. Conductors are to be chosen from Tables 310-16 through 310-19 or in accordance with Section 310-15. Requirements are noted for multimotor equipment.

440-7 Highest Rated (Largest) Motor

440-8 Single Machine

A refrigerating system or air-conditioning system can be considered in accordance with Section 430-87, exception, and Section 430-112, exception.

B. Disconnecting Means

440-11 General

440-12 Rating and Interrupting Capacity

It must be at least 115% of the larger of the rated load or branch-circuit election circuit. If this rating comes out to be 100 hp or larger, use the provisions of Section 430-109, Exception No. 4. Information is given concerning equivalent horsepower and locked rotor current. Additional information is given concerning equivalent horsepower, combination loads, small motor compressors and switches.

440-13 Cord-Connected Equipment

The separable connector or plug and receptacle can be used as the disconnecting means.

440-14 Location

It must be within sight and readily accessible to the equipment. There are two exceptions.

C. Branch-Circuit Short-Circuit and Ground-Fault Protection

440-21 General

440-22 Application and Selection

(a) Rating or Setting for Individual Motor-Compressor. It must allow the motor to start. The *NEC*® therefore permits a rating or setting of up to 175% of the larger of the motor load or branch-circuit selection current. This can be increased to 225% if necessary. A 15 ampere rating is the least that can be required for branch-circuit, short-circuit and ground-fault protection.

(b) Rating or Setting for Equipment. It must also allow the equipment to start. If only one hermetic refrigerant motor compressor is used, conform to Section 422(a). If more than one hermetic refrigerant motor compressor is used, conform to Section 430-53 and the following:

1. If the hermetic refrigerant motor compressor is the largest load, use Section 440-22(a) for this plus the other loads.
2. If the hermetic refrigerant motor compressor is not the largest load, the sum is that of the rating of the motor compressor loads plus the value noted in Section 430-53(c)(4) or Section 240-3 for other loads, whichever is applicable. There are two exceptions. For a detailed analysis, refer to the *NEC*® and one of the many reference guides.

(c) If a maximum rating is indicated by the manufacturer, it should not be exceeded.

D. Branch-Circuit Conductors

440-31 General

440-32 Single Motor-Compressor

The rating cannot be less than 125% of the largest of the motor compressor rated load or branch-circuit selection current.

440-33 Motor-Compressor(s) with or without Additional Motor Loads

The ampacity must be the sum of the larger of the rated load or branch-circuit selection currents of all the motor-compressor loads plus full-load currents of the other motors plus 25% of the largest motor. There are two exceptions to this: one is for room air conditioners as per Part G and the other is for interlocked motors.

440-34 Combination Load

The capacity must be adequate for all other loads, plus the motor-compressor load as specified in Sections 430-32 and 440-33. An exception is also made for interlocked circuitry.

440-35 Multimotor and Combination-Load Equipment

It cannot be less than that marked on the equipment as per Section 440-4(b).

E. Controller for Motor-Compressors

440-41 Rating

Generally, use the nameplate ratings (refer to Sections 440-7 and 440-6). If more than one load is served, use the combined load rating [refer to Section 440-12(b)]. There are additional requirements.

F. Motor Compressor and Branch-Circuit Overload Protection

440-51 General

440-52 Application and Selection

(a) **Protection of Motor-Compressor.** The following methods can be used: (1) an overload relay that will trip at not more than 140% of motor full-load current; (2) a thermal protector integral with the motor compressor; (3) a fuse or inverse time circuit breaker responsive to motor current and rated not more than 125% of motor current; and (4) a protective system furnished with the unit or specified for the unit.

(b) Protection of Motor-Compressor Control Apparatus and Branch-Circuit Conductors. The same types of devices used in Section 440-52(a) are permitted. It can be the same device protecting the motor-compressor in accordance with Section 440-52(a).

440-53 Overload Relays

440-54 Motor-Compressors and Equipment on 15 or 20 Ampere Branch Circuits—Not Cord and Attachment Plug-Connected

440-55 Cord and Attachment Plug-Connected Motor-Compressors and Equipment on 15 or 20 Ampere Branch Circuits

G. Provisions for Room Air Conditioners

440-60 General

440-61 Grounding

440-62 Branch-Circuit Requirements

The branch circuit can only have a load of 80% of its rating for a cord- and plug-connected air-conditioning unit, if no other loads are on the circuit. If other loads are on the circuit, then the load of the cord- and plug-connected air-conditioning unit cannot be more than 50% of the rating of the branch circuit.

440-63 Disconnecting Means

The attachment plug can be used for a single-phase unit at 250 Volts or less if the manual control on the unit is readily accessible and within 6 ft. of the floor or if a manual switch is readily accessible within sight of the unit.

440-64 Supply Cords

The length of the cord cannot be larger than 10 ft for 120 V or 6 ft. for 208 to 240 V.

The following article titles are listed for reference: **Article 445**—Generators: **Article 450**—Transformers and Transformer Vaults (Including Secondary Ties); **Article 455**—Phase Converters; **Article 460**—Capacitors; **Article 470**—Resistors and Reactors; **Article 480**—Storage Batteries.

5

SPECIAL OCCUPANCIES

ARTICLE 500
HAZARDOUS (CLASSIFIED) LOCATIONS

500-1 Scope—Articles 500 through 505

Articles 500 through 505 describe the electrical and electronic equipment and wiring requirements in locations where a fire or explosion hazard can exist because of flammable gases or vapors, flammable liquids, combustible dust, or ignitible fibers or flyings.

500-2 Location and General Requirements

The classifications are arrived at as a result of the items noted above and their properties. Modifications are noted in Articles 500 through 505. Additional explanatory fine-print notes are indicated.

(a) Protection Techniques. This is a new paragraph and describes protection techniques for the various hazardous (classified) locations.

Note: Space does not permit a complete description of this chapter. This is true of the entire *NEC*® but is again stated here. Only the highlights are noted. The reader should refer to the *NEC*® for complete descriptions.

500-3 Special Precaution

This section indicates the groupings of various air mixtures. Requirements are given for approval for class and properties, marking and temperature. If Article 505 is used there is a requirement for using a qualified registered professional engineer. Additional information is given in the form of fine print notes.

500-4 Specific Occupancies

These are covered in Articles 510 through 517 and include garages, aircraft hangars, gasoline dispensing and service stations, bulk storage plants, finishing processes, and health care facilities.

500-5 Class I Locations

Areas where flammable gases or vapors are in the air in quantities that can cause explosive or ignitible mixtures are classified as Class I. They are further broken down into two divisions:

(a) Class I, Division 1. This is where ignitible concentrations of gases or vapors can exist normally; where they may exist because of maintenance or leakage; or where they might exist because of breakdowns or faulty operations.

(b) Class I, Division 2. This is where volatile flammable liquids or flammable gases are used but are normally in closed containers or systems and can escape only as a result of a breakdown or rupture; where ignitible concentration of gases or vapors normally cannot occur because of mechanical ventilation; where an area is next to a Class I, Division 1, location and some leakage between areas occurs.

500-6 Class II Locations

These are hazardous because of combustible dust. They are also broken down into two divisions.

(a) Class II, Division 1. This is where combustible dust is in the air and can ignite under normal conditions due to its concentration, or where a breakdown of equipment can cause these concentrations and an ignitible source, or the dust may be electrically conductive in hazardous quantities.

(b) Class II, Division 2. This is where the concentrations of the dust are normally not high enough to ignite or explode and will not interfere with the operation of electrical equipment or other equipment; but can be in the air because of occasional breakdown of handling of processing equipment and can be ignited by electrical equipment or other equipment or prevent safe dissipation of heat from it.

500-7 Class III Locations

Areas that have ignitible fibers or flyings in a concentration not likely to cause ignition are classified as Class III. They are also further broken down into two divisions.

(a) Class III, Division 1. Areas where the fibers or flyings are handled, manufactured, or used.

(b) Class III, Division 2. Areas where the fibers or flyings are stored or handled. An exception is made in the process of manufacture.

ARTICLE 501
CLASS I LOCATIONS

501-1 General

This article refers to those areas classified as Class I.

501-2 Transformers and Capacitors

(a) Class I, Division 1

1. *Containing liquid that will burn.* They must be installed in an approved vault (Sections 450-41 through 450-48) with no door or opening to the Class I, Division 1 locations, with proper ventilation and the vents going outside the buildings and being large enough to relieve an explosion. The ducts for the vents must be constructed of reinforced concrete.
2. *Not containing liquid that will burn.* These must be installed in a vault as noted in (a)(1) or be approved for a Class I location.

(b) Class I, Division 2. In these locations they must be in accordance with Sections 450-21 through 450-27.

501-3 Meters, Instruments, and Relays

(a) Class I, Division 1. They must be in Class I, Division 1 enclosures.

(b) Class I, Division 2

1. *Contacts.* Make-or-break contacts, switches, circuit breakers, etc., must have an approved Class I, Division 1, enclosure. Four exceptions are: immersed in oil, in a hermetically sealed chamber, in nonincendive circuits, and where part of a listed nonincendive component.
2. *Resistors and similar equipment.* These must have an approved Class I, Division 1, enclosure. An exception is made if the exposed surface does not exceed 80% of the ignition temperature and there are no make-or-break or sliding contacts. The exception doesn't apply to thermionic tubes.
3. *Without make-or-break contacts.* General-purpose enclosures can be used for this category for transformer windings, impedence coils, solenoids, and other windings.
4. *General-purpose assemblies.* If an assembly has items that can use general-purpose enclosures [noted in (b)(1), (b)(2), and (b)(3)], a single, general-purpose enclosure can be used. If equipment noted in (b)(2) is used, the maximum surface temperature must be noted.
5. *Fuses.* Fuses for instrument circuits not subject to overloading can be mounted in general-purpose enclosures if approved in (b)(1) through (b)(4), and if they are preceded by a switch in accordance with (b)(1).
6. *Connections.* Process control can use a cord, plug, and receptacle if (a) an approved Class I, Division 1 switch is also used; (b) current is not greater than 3 A at 120 V; (c) cord is not longer than 3 ft., is rated for extra hard usage or hard usage depending on location, and uses locking and grounding-type plug and receptacle; and (d) a warning is posted against unplugging under load.

501-4 Wiring Methods

(a) Class I, Division 1. Only threaded rigid metal conduit, threaded steel intermediate conduit, or Type MI cable can be used. Boxes and fittings must be explosion proof and joints threaded. At least five threads must be engaged. MI cable has to be installed to prevent stress at the terminations. There are three exceptions.

(b) Class I, Division 2. Methods permitted are threaded rigid metal conduit; threaded steel intermediate metal conduit; enclosed gasketed busways; enclosed gasketed wireways; Type MI, MC, MV, or TC cable with approved terminations; Type PLTC cable. Explosion-proof boxes or fittings are not required. Additional requirements and exceptions are noted.

501-5 Sealing and Drainage

Seals and drainage must be provided and approved items must be used.

(a) Conduit Seals, Class I, Division 1

1. A seal must be placed in each conduit run which enters the enclosure for a switch, circuit breaker, relay, resistor, fuse, or any item that can cause arcing, sparking, or high temperatures. The seal must be within 18 in. of the enclosure. There is a listing of the enclosures or fittings that are allowed between the seal fitting and enclosure. The conduit bodies cannot be larger than the largest trade size of the conduits. There is an exception for conduit 1½ in. or smaller under special conditions.
2. If the conduit is 2 in. or larger, the seal must also be placed within 18 in. of enclosures containing terminals, taps, or splices.
3. If a run of conduit between two enclosures is not more than 36 in., one seal not more than 18 in. from an enclosure is acceptable.
4. A seal is required where a conduit run leaves a Class I, Division 1 location. The sealing fitting be within 10 ft. of the boundary on either side. The conduit must not have a fitting, union, coupling, or box between the seal and point of leaving the Class I, Division 1 location.

(b) Conduit Seals, Class I, Division 2

1. Seals are required for connection to explosion-proof enclosures and must comply with (a)(1) through (a)(3) above.
2. A seal is required when the conduit run leaves this area to an unclassified area. This paragraph describes additional requirements. There are four exceptions.

(c) Class I, Divisions 1 and 2. These are general rules.

1. The fittings must be approved for Class I locations.
2. The compound must be approved.
3. The thickness of the compound must be at least the trade size of the conduit and not less than ⅝ in.
4. Splices and taps cannot be made in fittings that are used for sealing.
5. Assemblies must be approved for Class I locations.
6. Conductor cross-sectional area in a seal cannot be more than 25% of a conduit of the same size unless approved for a higher fill.

(d) Cable Seals, Class I, Division 1

(e) Cable Seals, Class I, Division 2

(f) Drainage. In certain cases drainage of accumulated liquid is permitted and even required. Special requirements must be followed.

501-6 Switches, Circuit Breakers, Motor Controllers, and Fuses

(a) Class I, Division 1. Must be in an enclosure and the entire assembly must be approved for Class I locations.

(b) Class I, Division 2

1. Must be in enclosures approved for Class I, Division 1, locations unless a general-purpose enclosure is used and it is hermetically sealed, or has oil-immersed make-or-break contacts with special requirements, the device is solid state without contacts, or current interruption occurs in a factory-sealed explosion-proof chamber approved for the location.
2. If not intended to interrupt current, a general-purpose enclosure can be used with a disconnecting or isolating switch for transformers or capacitor banks.
3. Standard plug or cartridge fuses can be used in certain cases, but they must have enclosures approved for the location. If they are oil immersed, they can be placed in general-purpose enclosures.
4. If a distribution panel contains not more than 10 fuses or circuit breakers that are not to be used as switches, the fuses are enclosed, and the circuits are used only for fixed

lighting within one room or area, a general-purpose enclosure can be used.

5. Within lighting fixtures, approved cartridge fuses are permitted as supplementary protection.

501-7 Control Transformers and Resistors

Transformers, impedance coils, and resistors are covered by this section.

(a) Class I, Division 1. In this location they must have approved Class I, Division 1 enclosures.

(b) Class I, Division 2. In this location switching mechanisms must be in accordance with Section 501-6(b). General-purpose enclosures can be used for coils and windings. Resistors must have enclosures approved for Class I locations. If the resistor is fixed and the maximum operating temperature is not more than 80% of the ignition temperature of the gas or vapor, a general-purpose enclosure can be used.

501-8 Motors and Generators

(a) Class I, Division 1. In this location they must be either approved for Class I, Division 1 locations or be totally enclosed with a positive pressure ventilation (the arrangement being that the device cannot start until the ventilation has started and 10 volumes of air have been purged from the enclosure) or be totally enclosed filled with inert gas.

(b) Class I, Division 2. In this location you can have open or nonexplosion-proof enclosed motors if they do not have brushes, switching mechanisms, or arc-producing devices. If they do have brushes, switching mechanisms, or arc-producing devices, the motor or generator must be approved for Class I, Division 1, locations or the devices must be in Class I, Division 1 approved enclosures as per Section 501-3(b). A requirement is noted for the surface temperature of space heaters used to prevent condensation.

501-9 Lighting Fixtures

(a) Class I, Division 1. The entire fixture assembly must be approved for a Class I, Division 1 location and be protected against physical damage. Pendant fixtures must be wired and hung through rigid metal conduit or threaded steel intermediate conduit. Stems longer than 12 in. must have additional lateral bracing not more than 12 in. from the end of the stem or a flexible fitting approved for a Class I, Division 1 location. All boxes, fittings, etc., must be approved for Class I, Division 1 locations.

(b) Class I, Division 2. Portable lighting equipment must comply with Class I, Division 1 requirements. There is an exception when it is mounted on a moveable stand and connected by flexible cord. Fixed lighting must be protected against physical damage. If there is danger of falling sparks or hot metal igniting vapors or gases, proper enclosures must be provided. If the operating temperature will exceed 80% ofthe ignition temperature of the gases or vapors, they must comply with Section 501-9(a)(1) or be of the type that has been tested to determine the temperature range or operating temperature marked. Pendant fixtures have the same rules as Class I, Division 1 locations. Switches must be in accordance with Section 501-6(b)(1). Starting equipment for electric-discharge lamps must be in accordance with Section 501-7(b) unless it is part of a "thermally protected fluorescent lamp ballast" and the lighting fixture has the approval for this location.

501-10 Utilization Equipment

501-11 Flexible Cords Class I, Divisions 1 and 2

These are permitted only to connect the portable equipment to the fixed part of the system. They can also be used under certain conditions where a greater degree of movement is needed. As such, they must be approved for extra hard usage, contain a grounding conductor, be connected to the supply conductors or terminals in an approved way, ensure that there is no tension applied, and be provided with seals where necessary. Exceptions are made in Sections 501-3(b)(6) and 501-4(b). There is additional information concerning electrical submersible pumps and electric mixers.

501-12 Receptacles and Attachment Plugs, Class I, Divisions 1 and 2

Must be of the grounding type and approved for Class I locations except as provided in Section 501-3(b)(6).

501-13 Conductor Insulation, Class I, Divisions 1 and 2

If the conductors come in contact with vapors or gases, the insulation must be approved for this use or have protection provided by a lead sheath or another approved sheath.

501-14 Signaling, Alarm, Remote-Control, and Communication Systems

501-15 Live Parts, Class I, Divisions 1 and 2

In these locations, live parts are prohibited.

501-16 Grounding, Class I, Divisions 1 and 2

Wiring and equipment must be grounded in accordance with Article 250. Bonding jumpers with proper fitting or other approved means of bonding must be used. Locknut bushings and double locknut types of contacts are not acceptable. The means of bonding applies to all raceways, fittings, boxes, enclosures, etc., between Class I locations and the grounding point for service equipment or separately derived systems. If flexible conduit is used it must have internal or external bonding jumpers in parallel with each conduit. It must also comply with the requirements of Section 250-79 where flexible metal or liquidtight flexible metal conduit is used. There are exceptions.

501-17 Surge Protection, Class I, Divisions 1 and 2

501-18 Multiwire Branch Circuits

ARTICLE 502
CLASS II LOCATIONS

502-1 General

This article refers to areas classified as Class II. There is a definition of dust-ignition-proof. Note that "explosion-proof" and "dust-proof" are not the same; where dust-proof equipment is required, explosion-proof equipment cannot be required and used unless approved for Class II locations.

502-2 Transformers and Capacitors

(a) Class II, Division 1

1. *Containing liquid that will burn.* They must be installed in approved vaults outside the Division I location with self-closing fire doors, and have proper ventilation and pressure relief.

2. *Not containing liquid that will burn.* They must be installed in a vault as per Sections 450-41 through 450-48 or have the entire assembly approved for a Class II location.

3. If the area has a metal dust such as magnesium, aluminum, etc., you cannot install a transformer or capacitor.

(b) Class II, Division 2

1. *Containing liquid that will burn.* Must be installed in vaults as per Sections 450-41 through 450-48.

2. *Containing Askarel.* If it is larger than 25 kVA, it must have a pressure relief vent, have a way of absorbing gases from inside, and have an air space of at least 6 in. between the case and any combustible material.

3. *Dry-type transformers.* Must be in vaults or not operate over 600 V, with the windings and terminals enclosed without ventilating or other openings.

502-4 Wiring Methods

(a) Class II, Division 1. Only threaded rigid metal conduit, threaded steel intermediate metal conduit, Type MI, or Type SI cable can be used. Boxes and fittings can have no openings and have threaded bosses so that no dust can come in. If used where the dust is combustible electrically conductive, they must be approved for Class II locations. If a flexible connection must be used, it has to be either a dust-tight flexible connector, liquidtight flexible metal conduit, liquidtight nonmetallic conduit, or a flexible cord for extra-hard usage and bushed fittings. The flexible cord must comply with section 502-12. There is an exception for certain industrial establishments.

(b) Class II, Division 2. Methods permitted are rigid metal conduit, intermediate metal conduit, electrical metal tubing, dust-tight wireways, or Type MI or MC cable that have approved termination fittings or Type PLTC, MC, ITC, or TC cable under certain conditions. There is one exception. Also, wireways, fittings, and boxes must minimize the entrance of dust by the use of close-fitting covers and no openings. The requirements for flexible connections are the same as for Class II, Division 1.

502-5 Sealing, Class II, Divisions 1 and 2

If a raceway connects two enclosures, one being dust-ignition-proof, the other not, provision must be made to prevent dust from getting into the dust-ignition-proof enclosures. This can be done with seals, a minimum of 10 ft of horizontal raceway, or a minimum of 5 ft of vertical raceway. If the dust-ignition-proof enclosure is connected to one in an unclassified area, a seal is not required.

502-6 Switches, Circuit Breakers, Motor Controllers, and Fuses

(a) Class II, Division 1. Their enclosures must be dust-ignition-proof. If the dust is a hazardous metal (magnesium, etc.), the enclosure must be approved for this specific type of location. If they are not installed where there is electrically conductive dust, are not to interrupt current, and have no fuses, the enclosure only has to be tight so as to keep the entrance of dust to a minimum and have no openings, as well as close-fitting or telescoping covers.

(b) Class II, Division 2. The enclosures only have to be dust-tight.

502-7 Control Transformers and Resistors

(a) Class II, Division 1. The enclosures must be dust-ignition-proof approved for a Class II location. If they are installed in a location with hazardous metal dust (magnesium, etc.), the enclosure must be approved for this specific type of location.

(b) Class II, Division 2. In these locations, the switching mechanisms must have a dust-tight enclosure. If they are in the same location as the switching mechanism, they must have tight metal enclosures with no openings. Resistors must have approved Class II enclosures unless their temperature is not higher than 248°F and they are either nonadjustable or part of an automatic timing system. In that case the enclosures have only to be tight with no openings.

502-8 Motors and Generators

(a) Class II, Division 1. They must be either totally enclosed pipe-ventilated and meet the temperature restrictions of Section 502-1 or be approved for locations classified as Class II, Division 1.

(b) Class II, Division 2. They must be either totally enclosed pipe-ventilated, totally enclosed fan-cooled, totally enclosed water–air-cooled, totally enclosed nonventilated or dust-ignition-proof with full load external temperature in accordance with Section 500-3(f). There are exceptions for these types of motors if the authority having jurisdiction feels that the machines are easily cleaned and maintained and there will not be a large accumulation of dust.

502-9 Ventilating Piping

The piping must (1) be metal at least 0.021 in. thick or of noncombustible material of equal substance, (2) be screened at the outside ends, (3) lead to the outside of the building and have clean air available, and (4) be protected against corrosion and physical damage. In addition, it must comply with:

(a) Class II, Division 1. The pipes must be entirely dust-tight. They must also be either riveted and soldered, bolted and soldered, welded, or be dust-tight by another effective means.

(b) Class II, Division 2. The pipes must be tight enough to prevent the entrance of dust in appreciable quantities.

502-10 Utilization Equipment

(a) Class II, Division 1. Must be approved for Class II locations. If hazardous dust is present, it must also be approved for that location.

(b) Class II, Division 2. Heaters must be approved for Class II locations. There is one exception. Motors must be in accordance with Section 502-8(b). Dust-tight enclosures are required for circuit breakers, switches, and fuses. Section 502-7 is applied to transformers, impedance coils, and resistors.

502-11 Lighting Fixtures

(a) Class II, Division 1. The fixtures must be approved for Class II locations and additionally for hazardous metal dust if so located. They must be protected against physical damage. Pendant fixtures must be suspended using threaded rigid metal conduit, threaded steel intermediate metal conduit, or approved chains or other means. If the stem is longer than 12 in., it must be braced within 12 in. from the lower end or have a flexible fitting not more than 12 in. from the upper end. If conduit is not used for the wiring, the cord must be approved for hard usage and proper seals must be used. All boxes, fittings, etc., must be approved for Class II locations.

(b) Class II, Division 2. Portable lighting equipment must be approved for Class II locations. A fixed lighting fixture must only minimize the collection of dust and prevent the escape of sparks. It must also have maximum wattage permitted for temperatures in accordance with Section 500-3(d). Fixtures in these locations must also be protected against physical damage. Requirements for pendant fixtures in Division 2 are basically the same as for Division 1. One difference is that seals are not required. The requirements of Section 502-7(b) apply here for starters and control equipment for electric-discharge lamps.

502-12 Flexible Cords, Class II, Divisions 1 and 2

502-13 Receptacles and Attachment Plugs

(a) Class II, Division 1. Must be approved for Class II locations and have a grounding conductor connection.

(b) Class II, Division 2. Must also have a grounding conductor connection. They cannot permit the supply circuit to open with exposed live parts.

502-14 Signaling, Alarm, Remote-Control, and Communication Systems, Meters, Instruments and Relays

502-15 Live Parts, Class II, Divisions 1 and 2

Live parts are not permitted to be exposed.

502-16 Grounding, Class II, Divisions 1 and 2

Ground, in accordance with Article 250, all wiring and equipment.

(a) Bonding jumpers with proper fittings or other approved means must be used. Locknut bushings and double locknut contacts alone are not permitted. The means of bonding applies to all raceways, fittings, boxes, enclosures, etc., between Class II locations and the grounding point for service equipment or separately derived systems. There is an exception.

(b) If flexible conduit is used, it must have internal or external bonding jumpers in parallel with each conduit. It must also comply with the requirements of Section 250-79. There is an exception.

502-17 Surge Protection, Class II, Divisions 1 and 2

502-18 Multiwire Branch Circuits

ARTICLE 503
CLASS III LOCATIONS

503-1 General

Maximum surface temperatures for equipment permitted under operating conditions are 329°F for normal operation or 248°F if the equipment can be overloaded.

503-2 Transformers and Capacitors, Class II, Divisions 1 and 2

Must be in accordance with Section 502-2(b)

503-3 Wiring Methods

(a) Class III, Division 1. Use only rigid metal conduit, rigid nonmetallic conduit, intermediate metal conduit, electrical metallic tubing, dust-tight wireways or Type MI or MC cables with approved fittings. Fittings and boxes must be dust tight. Refer to this section for the requirements for flexible connections.

(b) Class III, Division 2. Same requirements as Division 1. An exception is made for storage areas with no machinery, where open wire on insulators can be used complying with Article 320, specifically Section 320-14.

503-4 Switches, Circuit Breakers, Motor Controllers and Fuses Class III, Divisions 1 and 2

These devices must have dust-tight metal enclosures.

503-5 Control Transformers and Resistors, Class III, Divisions 1 and 2

Must be in dust-tight enclosures complying with temperature limitations in Section 503-1.

503-6 Motors and Generators, Class III, Divisions 1 and 2

(a) Class III, Division 1. Must be totally enclosed pipe ventilated, totally enclosed nonventilated, or totally enclosed fan cooled. There is an exception where the authority having jurisdiction will permit it.

503-7 Ventilating Piping, Class III, Divisions 1 and 2

Must be metal at least 0.021 in. thick or equal noncombustible material, have physical damage protection, go directly outside to clean air, and be screened. The pipes must be tight enough to prevent "appreciable quantities" of fibers or flyings from getting in and sparks getting out.

503-8 Utilization Equipment, Class III, Divisions 1 and 2

Heaters: approved for Class III locations. Motors: Section 503-6. Switches, circuit breakers, motor controllers, and fuses: Section 503-4.

503-9 Lighting Fixtures, Class III, Divisions 1 and 2

Protect against physical damage. Prevent entrance of fibers or flyings or escape of sparks. Pendant stems must be threaded rigid metal conduit, threaded intermediate metal conduit, or threaded metal tubing. Chains with approved fittings can be used for suspending fixtures. The same rules apply for bracing stems over 12 in. long as Class II. Portable lamps must have handles, guards, and the lampholders have no switches, exposed parts, or receptacles.

503-10 Flexible Cords, Class III, Divisions 1 and 2

They must be approved for extra-hard usage, have a grounding conductor, be connected in an approved manner, be supported so as to prevent tension on the connections, and have provisions to prevent fibers or flyings from entering the boxes or fittings.

503-11 Receptacles and Attachment Plugs, Class III, Divisions 1 and 2

They must be of the grounding type, and designed to prevent the escape of sparks or molten particles and minimize the entry of fibers or flyings. An exception is provided where permitted by the authority having jurisdiction.

503-12 Signaling, Alarm, Remote-Control, and Local Loud Speaker Intercommunication Systems, Class III, Divisions 1 and 2

503-13 Electric Cranes, Hoists, and Similar Equipment, Class III, Divisions 1 and 2

503-14 Storage-Battery Charging Equipment, Class III, Divisions 1 and 2

503-15 Live Parts, Class III, Divisions 1 and 2

No exposed live parts permitted.

503-16 Grounding, Class III, Divisions 1 and 2

As per Article 250. Also, the same requirements hold true for bonding and grounding as in Sections 501-16, and 502-16.

ARTICLE 504
INTRINSICALLY SAFE SYSTEMS

ARTICLE 505
CLASS I, ZONE 0, 1, AND 2 LOCATIONS

If this article is used, the equipment selection, wiring, and area classification must be under the supervision of a qualified registered professional engineer.

505-2 General Requirements

505-5 Grouping and Classification

505-7 Zone Classification

505-10 Listing and Marking

505-15 Wiring Methods

505-20 Equipment

505-25 Grounding and Bonding

ARTICLE 510 HAZARDOUS (CLASSIFIED) LOCATIONS-SPECIFIC

Articles 511 through 517 cover specific locations that might be hazardous. The following article titles are listed for reference: **Article 511**—Commercial Garages, Repair and Storage; **Article 513**—Aircraft Hangars; **Article 514**—Gasoline Dispensing and Service Stations; **Article 515**—Bulk Storage Plants; **Article 516**—Spray Application, Dipping and Coating Processes.

ARTICLE 517 HEALTH CARE FACILITIES

A. General

517-1 Scope

Electrical construction and installation criteria for health care facilities are covered in this article. Performance, maintenance, and testing criteria are covered in NFPA 99 and other appropriate health care documents. Veterinary facilities are not covered. The reader is referred to NFPA 99.

517-2 General

Parts B and C are intended to be applied to both single-function buildings and to applicable portions of multifunction buildings.

The example used is a doctor's examining room in a residential custodial care facility having to comply with Section 517-10.

517-3 Definitions

Many definitions are listed in this section that are not covered in Article 100. These are specific to health care facilities. The reader should refer to these in the Code.

B. Wiring Design and Protection

517-10 Applicability

Generally, this part applies to all health care facilities. There are two exceptions: (1) business offices, corridors, waiting rooms, and similar areas in clinics, medical and dental offices, and outpatient clinics, and (2) if wired in accordance with Chapters 1, 2, 3, and 4 of the Code, then patient sleeping areas in nursing homes and residential care facilities are not included. The second exception does not apply where patients will be on life support systems, have invasive procedures and be on line-operated electromedical devices.

517-11 General Installation/Construction Criteria

It is important that the system installed maintain low potential differences between exposed conductive surfaces which are likely to become energized and the patient could contact.

517-12 Wiring Methods

All the requirements of Chapters 1 through 4 must be followed unless they are changed in this article.

517-13 Grounding of Receptacles and Fixed Electric Equipment

(a) Patient Care Areas. If operated over 100 V, the grounding terminal of all receptacles and noncurrent-carryingareas of fixed equipment that a person can contact and may become energized must be grounded using an insulated copper conductor. It must be sized according to Table 250-95 and be installed in a metal raceway with the branch-circuit conductors supplying the equipment or receptacle. There are three exceptions.

(b) Methods. Branch circuits serving patient care areas must have a ground path for fault current by being installed in certain metal raceways or cable assemblies where the outer jacket is an approved grounding means. This is in addition to all other requirements of this section.

517-14 Panelboard Bonding

The equipment grounding busses of all panelboards serving the same patient vicinity must be bonded together. This is to be done using a continuous insulated copper conductor No. 10 AWG or larger.

517-16 Receptacles with Insulated Grounding Terminals

517-17 Ground-Fault Protection

(a) Feeders. If ground-fault protection is used at the service disconnect as per Section 230-95 or 215-10, an additional step is needed at the next level of feeder away from the service. There are three areas listed where the additional level of ground-fault protection cannot be installed.

(b) Selectivity. The system must be fully selective. Six-cycle separation between the levels is required.

(c) Testing.

517-18 General Care Areas

(a) Patient Bed Location Branch Circuits. Each patient bed location must have at least two branch circuits, at least one of them must originate from the normal system and one from the emergency system. All branch circuits on the normal system must come from the same panel-board. There are two exceptions to this.

(b) Patient Bed Location Receptacles. Each location must have at least four receptacles, either single, duplex, or a combination. They must all be hospital grade and grounded with an insulated copper conductor. The requirement that the receptacle must be hospital grade listed went into effect January 1, 1993. Use Table 250-95. There are two exceptions to this.

(c) Pediatric Locations. The 15- and 20-A receptacles must be tamper resistant. A definition of tamper resistant is noted. There is an exception.

517-19 Critical Care Areas

(a) Patient Bed Location Branch Circuits. Each patient bed location must have at least two branch circuits, at least one from the normal system and one from the emergency system. One circuit on the emergency system that supplies no other bed location must be available at each bed location. All branch circuits from the normal system must come from one panelboard. The emergency receptacles must be identified and have the panelboard and circuit number of its supply noted. There are two exceptions.

(b) Patient Bed Location Receptacles. Each location must have at least six receptacles, either single, duplex, or a combination of both. At least one must be connected to the normal system. They must all be hospital grade and grounded with an insulated copper conductor to the reference grounding point.

(c) Patient Vicinity Grounding and Bonding (optional). A patient equipment grounding point is allowed in the patient vicinity. It can have jacks approved for that purpose. All the grounding terminals on the receptacles must be connected to the patient equipment grounding point if supplied with an equipment bonding jumper no smaller than No 10. AWG. This jumper can be looped or run centrically.

(d) Panelboard Grounding. Grounding of panelboards and switchboards must be accomplished if a grounded distribution system is used and metal feeder raceway or Type MC or MI cable is used. There are three acceptable means noted for use at each termination or junction point in the system.

(e) Additional Protective Techniques in Critical Care Areas (optional). Isolated power systems are permitted.

(f) Isolated Power System Grounding.

(g) Special-Purpose Receptacle Grounding.

517-20 Wet Locations

(a) In a wet location two methods can be used to protect personnel. Either a ground-fault circuit interrupter can be used if power interruption is acceptable or an isolated power supply can be used. There is an exception.

(b) Isolated power supplies must conform to Section 517-160.

517-21 Ground-Fault Circuit Protection for Personnel

C. Essential Electrical System

517-25[x] Scope

This system is needed to provide light and power for systems essential to life safety and orderly cessation of procedures during a power failure of the normal electrical supply. The types of structures and services are listed in the Code. The reader is also referred to NFPA 99-1993 (ANSI) for further information and for performance, maintenance, and testing requirements.

517-30 Essential Electrical Systems for Hospitals

(a) Applicability.

(b)[x] General. The system is comprised of two separate systems, the emergency system and the equipment system. The emergency systems has a life safety branch and a critical branch. The equipment system supplies major equipment. The number of transfer switches depends on many design considerations. One switch can be used for a system with a maximum demand of 150 kVA as indicated in Diagram 517-30(c). Loads not noted in Sections 517-33 and 517-34 must have their own transfer switch and must not transfer if it overloads the generator and must automatically shed if the generating equipment becomes overloaded.

(c) Wiring Requirements. The life safety and critical branch must be kept separate from all other wiring and equipment. The wiring of the emergency system must be mechanically protected in metal raceways (there are six exceptions to this). When isolated power systems are used in anesthetizing locations or special environments, each must be supplied by a separate circuit. There are four exceptions.

(d) Capacity of Systems. The system must be able to supply the entire load. There is now a statement on demand calculations.

517-31[x] Emergency System

This is divided into two required branches, the life safety and the critical. This system must be automatically restored to power within 10 seconds of the loss of the normal supply.

517-32[x] Life Safety Branch

The Code lists all the items that are required to be on the life safety branch. They are basically illumination of means of egress, exit signs, alarm and alerting systems, communication systems, generator set location, and elevators. The reader is referred to the Code for the exact list. No other function other than those listed can be connected to the life safety branch.

517-33^x Critical Branch

The Code lists all the items that are required to be on the critical branch. They are basically task illumination and selected receptacles for the isolated power system, anesthetizing locations, patient care areas, additional specialized areas, nurse call systems, blood, bone and tissue bank, telephone equipment room and closets, special areas, and special circuits. The critical branch can be divided into two or more branches. The reader is referred to the Code for the exact list.

517-34^x Equipment Source Connection to Alternate Power Source

This section describes what equipment must be placed on the equipment system. It also describes how the system is energized (i.e., automatic, delayed automatic, and manually). The reader is referred to the Code for the exact requirements.

517-35 Sources of Power

(a)^x **Two Independent Sources of Power.** There must be a normal source of power and an alternate source of power for use when the normal source is interrupted.

(b)^x **Alternate Source of Power.** This must be located on the premises and be a generator driven by a prime mover. There is an exception to this when the normal source is a generator. In that case the alternate source can be either another generator or an outside utility company.

(c) Location of Essential Electrical System Components.

517-40 Essential Electrical Systems for Nursing Homes and Limited Care Facilities

(a) Applicability.

(b) Inpatient Hospital Care Facilities. If the facility provides inpatient hospital care, it must comply with the requirements for hospitals.

(c) Facilities Contiguous with Hospitals. These can have their systems supplied by the hospital. The reader is again referred to NFPA 99-1993 (ANSI) for performance, maintenance, and testing requirements.

517-41 Essential Electrical Systems

(a)^x **General.** The systems for nursing homes and limited care facilities must have two separate branches, the life safety branch and the critical branch.

(b)[x] **Transfer Switches.** The number of transfer switches depends on design. One transfer switch is permitted if the load is not more than 150 kVA and as shown in Diagram 517-41(c).

(c) Capacity of System. The system must be capable of supplying the entire load.

(d) Separation from Other Circuits. The life safety branch must be kept separate from all other wiring and equipment. There are three exceptions.

517-42[x] Automatic Connection to Life Safety Branch

The life safety branch must be automatically restored to power after 10 seconds of loss of the normal source of supply. This branch is referred to as the emergency system in NFPA 99-1993 (ANSI). The life safety branch basically supplies power for illumination of means of egress, exit signs, alarm and alerting systems, communication systems, dining and recreation areas, generator set location, and elevators. Refer to the Code for complete requirements. No other function other than those listed can be connected to the life safety branch.

517-43[x] Connection to Critical Branch

This section describes what must be connected to the critical branch and how it is to be energized from the essential electrical system (i.e., automatic, delayed automatic, or manual).

517-44 Sources of Power

(a)[x] **Two Independent Sources of Power.** There must be a normal source of power and an alternate source of power for use when the normal source is interrupted.

(b)[x] **Alternate Source of Power.** This must be located on the premises and be a generator driven by a prime mover. There is an exception to this when the normal source is a generator. In that case the alternate source can be either another generator or an outside utility company. There is also an exception where battery units may be used under certain conditions, which is a second exception for nursing homes or limited-care facilities meeting the requirements of Section 517-40(a) Exception.

(c) Location of Essential Electrical System Components.

517-45 Essential Electrical Systems for Ambulatory Health Care Centers

(a) Applicability.

(b) Connections. It must supply power for task illumination to bring a procedure to an orderly cessation for safety to life for the patient, all anesthesia and resuscitative equipment used where inhalation anesthetics are induced, and all electrical life support equipment.

(c) Alternative Source of Power. The three paragraphs below are extracted from NFPA 99.

(1) *Power Source

(2) *System Capacity

(3) *System Operation

517-50 Essential Electrical Systems for Clinics, Medical Office Buildings and Dental Offices, Outpatient Facilities, and Other Health Care Facilities Not Covered in Sections 517-30, 517-40, and 517-45

There are requirements for these facilities discussed.[x] They can be less restrictive depending on the occupancy and type of service rendered.

D. Inhalation Anesthetizing Locations

The reader is again referred to NFPA 99-1993 (ANSI).

517-60 Anesthetizing Location Classification

These are classified as hazardous (classified) or other-than-hazardous (classified).

517-61 Wiring and Equipment

517-62 Grounding

517-63 Grounded Power Systems in Anesthetizing Locations

517-64[x] Low-Voltage Equipment and Instruments

E. X-Ray Installations

517-71 Connection to Supply Circuit

517-72 Disconnecting Means

517-73 Rating of Supply Conductors and Overcurrent Protection

517-74 Control Circuit Conductors

517-75 Equipment Installations

517-76 Transformers and Capacitors

517-77 Installation of High-Tension X-Ray Cables

517-78 Guarding and Grounding

F. Communications, Signaling Systems, Data Systems, Fire Protective Signaling Systems, and Systems Less Than 120 Volts, Nominal

517-80 Patient Care Areas

517-81 Other Than Patient Care Areas

517-82 Signal Transmission between Appliances

G. Isolated Power Systems

517-160 Isolated Power Systems

(a) Installations
(b)[x] Line Isolation Monitor

The following article titles are listed for reference: **Article 518**—Places of Assembly; **Article 520**—Theaters, Audience Areas of Motion Picture and Television Studios, and Similar Locations; **Article 525**—Carnivals, Circuses, Fairs, and Similar Events; **Article 530**—Motion Picture and Television Studios and Similar Locations; **Article 540**—Motion Picture Projectors; **Article 545**—Manufactured Buildings; **Article 547**—Agricultural Buildings; **Article 550**—Mobile Homes, Manufactured Homes and Mobile Home Parks; **Article 551**—Recreational Vehicles and Recreational Vehicle Parks; **Article 552**—Park Trailers; **Article 553**—Floating Buildings; **Article 555**—Marinas and Boatyards.

6, 7, and 8

***Note:* Those articles in Chapters 6, 7, and 8 not discussed in their entirety are listed here with article titles for reference purposes.**

CHAPTER 6
SPECIAL EQUIPMENT

Article 600—Electric Signs and Outline Lighting; **Article 604**—Manufactured Wiring Systems; **Article 605**—Office Furnishings; **Article 610**—Cranes and Hoists; **Article 620**—Elevators, Dumbwaiters, Escalators, Moving Walks, Wheel Chair Lifts, and Stairway Chair Lifts; **Article 625**—Electric Vehicle Charging System Equipment; **Article 630**—Electric Welders; **Article 640**—Sound-Recording and Similar Equipment.

ARTICLE 645
INFORMATION TECHNOLOGY EQUIPMENT

645-1 Scope

This article covers all types of wiring and grounding (i.e., power supply, equipment interconnect, etc.) in information technology equipment rooms. The reader can refer to this article as well as NFPA 75-1995 ANSI Protection of Electronic/Data

Processing Equipment (whose rules this room must comply with) for further information.

645-2 Special Requirements for Information Technology Equipment Room

Article 645 will apply if all the following are met: (1) Disconnecting means are provided that are in accordance with Section 645-10; (2) a separate HVAC system is provided for the room, or another system in another area is used if fire/smoke dampers which operate from smoke detectors and the disconnecting means in the room are provided in the ducts at the point of entry into the room; (3) the electronic computer/data processing equipment is listed; (4) the room is occupied only by the personnel needed to operate and maintain the equipment; (5) fire-resistant-rated walls and protected openings are used to separate the area from other areas; and (6) all applicable building codes are complied with.

645-5 Supply Circuits and Interconnecting Cables

(a) The ampacity of the branch-circuit conductors must be at least 125% of the total connected load.

(b) The data processing system can be connected to branch circuits using computer/data processing cables and attachment plug caps, flexible cords and attachment plug caps, or cord set assemblies. They must be listed for that purpose. If a cord set assembly is used on the floor, it must be protected from physical damage.

(c) Separate units can be interconnected with cables and cable assemblies. They must be listed for this use and protected from physical damage if installed on the floor.

(d) There are a number of rules for circuits, cables, and devices under a raised floor. Cables and receptacles are allowed to be installed under a raised floor. The area under the floor has to be accessible and the floor must be of proper construction. Branch-circuit conductors must be installed in rigid or intermediate metal conduit, rigid nonmetallic conduit, electrical metallic tubing, metal wireway, surface metal raceway with a metal cover, flexible metal conduit, liquidtight flexible metal or nonmetallic conduit, Type MI cable, Type MC cable, or Type AC cable and installed in accordance with Section 300-11. The openings in the raised floor for cables have protection against abrasion to the cables and a means to minimize the debris that will fall onto the floor. The ventilating system under the floor is used only for the data processing equipment and area. Cables other than those listed above must be listed as type DP cable with adequate fire-resistance to be used under a raised floor. This became effective July 1, 1994. There are three exceptions to this.

(e) All cables and accessories that are part of or used for the equipment have to be properly secured in place.

645-6 Cables Not in Computer Room

All cables that go outside the room have to comply with the *NEC®*

645-7 Penetrations

Utilize Section 300-21 for rules applying to all penetrations of the fire-resistant separation of the room.

645-10 Disconnecting Means

Disconnecting means must be provided for all system equipment and HVAC equipment. This can be one central one or separate ones for different types of equipment. The disconnects have to be grouped together, identified, and be accessible at the principal exit. The exception is for those installations under Article 685.

645-11 Uninterruptible Power Supplies

These are commonly referred to as UPS. These systems have to comply with Section 645-10. The disconnecting means must disconnect the battery from its load. There are two exceptions.

645-15 Grounding

All exposed noncurrent carrying metallic parts of an electronic computer-data processing equipment system has to be grounded as noted in Article 250 or be double insulated. If the derived power system is in listed equipment and supplies power to systems through cable assemblies or receptacles which are part of the equipment, it is not considered a separately derived system as per Section 250-5(d). Exception No. 4 Section 250-74 is referred to for isolated grounding type receptacles.

645-16 Marking

Each piece of system equipment supplied by a branch circuit must have a manufacturer's nameplate. The nameplate must show the maximum rated load in amperes, voltage, and frequency.

The following article titles are listed for reference: **Article 650**—Pipe Organs; **Article 660**—X-Ray Equipment; **Article 665**—Induction and Dielectric Heating Equipment; **Article 668**—Electrolytic Cells; **Article 669**—Electroplating; **Article 670**—Industrial Machinery; **Article 675**—Electrically Driven or Controlled Irrigation Machines; **Article 680**—Swimming Pools, Fountains, and Similar Installations; **Article 685**—Integrated Electrical Systems, **Article 690**—Solar Photovoltaic Systems, **Article 695**—Fire Pumps.

CHAPTER 7
SPECIAL CONDITIONS

ARTICLE 700
EMERGENCY SYSTEMS

A. General

700-1 Scope

This article lays out the electrical safety requirements for the installation, operation, and maintenance of emergency systems. Emergency systems include all items necessary to deliver electricity to the load. Emergency systems are defined as those which are legally required by any governmental agency or specific code that has jurisdiction. They supply, distribute, and control electricity for power and illumination which is essential for safety to human life. The code goes into detail with specific types of occupancies and equipment. The reader is referred to other articles in this code or to other NFPA documents for additional information as follows:

NEC Article 517 Health Care Facilities for wiring and installation in health care facilities

NFPA 99-1993 for performance and maintenance in health care facilities

NFPA 101-1994 Life Safety Code for locations where the systems are essential to life safety

NFPA 110-1993 Emergency and Standby Power Systems for performance of these systems

These additional documents and articles may place other restrictions on the design of the emergency system for the specific application of the document or article.

700-2 Application of Other Articles

All applicable articles of the *NEC*® apply unless they are modified by this Article.

700-3 Equipment Approval

All equipment must be approved for this use.

700-4 Tests and Maintenance

(a) Conduct or Witness Test. The system must be tested at installation and periodically thereafter. The test must be conducted and witnessed by the authority having jurisdiction.

(b) Tested Periodically. To assure proper maintenance and operation the system must be tested on a schedule accepted by the authority having jurisdiction.

(c) Battery System Maintenance. Batteries, whether for systems or in conjunction with engines, must be maintained periodically.

(d) Written Record. A written record must be kept of all tests and maintenance.

(e) Testing under Load. Provisions must be made to test the system under the maximum anticipated load.

700-5 Capacity

(a) Capacity and Rating. The system must be able to supply the entire load operating together. It must also be able to withstand the maximum available fault current.

(b) Selective Load Pickup, Load Shedding, and Peak Load Shaving. This permits the system to be used for various levels of loads and peak shaving provided that it can supply the emergency circuits, the legally required standby circuits, and the optional standby circuits when required.

700-6 Transfer Equipment

The transfer equipment must be automatic. Interconnection of the normal and emergency sources must be prevented. Bypass switches for isolation of the transfer switch must be installed.

700-7 Signals

Audible and visual signal must be installed where practical to indicate the derangement of the emergency source, that the battery is carrying the load, the battery charger is not functioning, and there is a ground fault in solidly grounded wye systems of more than 150 V to ground and circuit protective devices rated more than 1000 A. There is additional information concerning the ground-fault devices.

700-8 Signs

B. Circuit Wiring

700-9 Wiring Emergency Systems

(a) Identification. All items in the system must be readily identified as part of the system.

(b) Wiring. The wiring of the system must be separate from the wiring of the normal system and cannot be in the same raceway, boxes, cabinets, cables, etc. There are five exceptions. The circuits must be designed and located to reduce to a minimum the hazards due to flooding, vandalism, icing, and fires.

(c) Fire Protection. Additional requirements are noted.

C. Sources of Power

700-12 General Requirements

The current must be available for use within 10 seconds after the loss of normal power. The supply source can be any of the following: storage batteries, generator set, separate service, uninterruptible power supply, connection ahead of the service disconnecting means, and unit equipment. Each one is covered in the code in detail as to specific requirements. Again it must be noted that some of these may not be permitted for specific occupancies as noted in other documents (i.e., health care facilities). The equipment must be designed and located so that hazards from flooding, vandalism, icing, and fires are minimized.

There are new requirements for buildings with certain sizes and equipment. There are also additional requirements for storage batteries, generator set ups, separate service, and unit equipment.

D. Emergency System Circuits for Lighting and Power

700-15 Loads on Emergency Branch Circuits

Only those appliances and lamps needed for emergency use can be connected to the emergency lighting circuits.

700-16 Emergency Illumination

700-17 Circuits for Emergency Lighting

700-18 Circuits for Emergency Power

E. Control—Emergency Lighting Circuits

700-20 Switch Requirements

700-21 Switch Location

700-22 Exterior Lights

F. Overcurrent Protection

700-25 Accessibility

700-26 Ground-Fault Protection of Equipment

This is not required for the alternate source. Ground fault indication is required in accordance with Section 700-7(d).

ARTICLE 701
LEGALLY REQUIRED STANDBY SYSTEMS

A. General

701-1 Scope

This article lays out the electrical safety requirements for the installation, operation, and maintenance of legally required standby power systems. This includes all equipment necessary to deliver power to the load. The reader is referred to NFPA 99-1993 (ANSI) Health Care Facilities for additional information, NFPA 110-1993 (ANSI) Emergency and Standby Power Systems for performance information, and to ANSI/IEEE 446-1987 Recommended Practice for Emergency and Standby Power Systems for Industrial and Commercial Applications for further information.

701-2 Legally Required Standby Systems

These are systems that have this classification set by any governmental agency having jurisdiction. They are intended to supply power automatically to selected loads other than emergency system loads if the normal source fails. They differ from emergency systems in that they are not required for direct safety to life. They do supply power where the interruption might create a hazard or hamper rescue or firefighting operations. Some examples are given in the code to help the reader.

701-3 Application of Other Articles

Applicable articles of the Code apply unless otherwise modified in this Article.

701-4 Equipment Approval

Items must be approved for this use.

701-5 Test and Maintenance for Legally Required Standby Systems

(a) Conduct or Witness Test. When the installation is complete the authority having jurisdiction must either conduct or witness a test.

(b) Tested Periodically. To assure proper maintenance and operation the system must be tested on a schedule accepted by the authority having jurisdiction.

(c) Battery Systems Maintenance. Periodic maintenance is required if the batteries are used for control, starting, or ignition of a prime mover.

(d) Written Record. A written record must be kept of all tests and maintenance.

(e) Testing under Load. Provisions must be made for testing under load.

701-6 Capacity and Rating

The system must be able to supply the entire load operating together. It must also be able to withstand the available fault current.

701-7 Transfer Equipment

The transfer equipment must be automatic. Interconnection of the normal and emergency sources must be prevented. Bypass switches for isolation of the transfer switch must be installed.

701-8 Signals

Audible and visual signals must be provided to indicate the derangement of the standby source, that the standby source is not carrying the load, and that the battery charger is not functioning.

701-9 Signs

The type and location of the system must be indicated on a sign at the service entrance. There is an exception for certain unit equipment. If the system is grounded to a remote grounding electrode a sign must be placed at the grounding location identifying all emergency and normal sources connected there.

B. Circuit Wiring

701-10 Wiring Legally Required Standby Systems

The wiring for this system can occupy the raceways, cables, boxes, cabinets, etc. as the general wiring.

C. Sources of Power

701-11 Legally Required Standby Power Systems

The current must be available for use within 60 seconds of the loss of the normal power. The supply system can be any one of the following: storage battery, generator set, uninterruptible power supply, separate service, connection ahead of the service disconnecting means, and unit equipment. Each is covered in the code in detail as to specific requirements. Again it must be noted that some of these may not be permitted for specific occupancies as noted in other documents (i.e., health care facilities).

D. Overcurrent Protection

701-15 Accessibility

701-16 Ground-Fault Protection of Equipment

This is not required for the alternate source.

ARTICLE 702
OPTIONAL STANDBY SYSTEMS

A. General

702-1 Scope

This article applies only to the installation and operation of optional standby systems.

702-2 Optional Standby Systems

These systems protect business or property where life safety is not dependent on the system. Power can be supplied either automatically or manually.

702-3 Application of Other Articles

Unless modified by this article, all provisions of the Code apply.

702-4 Equipment Approval

All equipment must be approved for this use.

702-5 Capacity and Rating

The system must be able to supply the entire load operating together. It must also be able to withstand the available fault current. The user can select the load which is connected to the system.

702-6 Transfer Equipment

If it is located on the load side of branch-circuit protection it can have supplementary overcurrent protection with an interrupting rating that can handle the generator available fault current.

702-7 Signals

Audible and visual signals must be provided to indicate the derangement of the standby source and that the standby source is carrying the load.

702-8 Signs

The type and location of the system must be indicated on a sign at the service entrance. There is an exception for certain unit equipment. If the system is grounded to a remote grounding electrode a sign must be placed at the grounding location identifying all emergency and normal sources connected there.

B. Circuit Wiring

702-9 Wiring Optional Standby Systems

The wiring for this system can occupy the raceways, cables, boxes, etc., as the general wiring.

The following article titles are listed for reference: **Article 705**—Interconnected Electric Power Production Sources; **Article 710**—Over 600 Volts, Nominal—General; **Article 720**—Circuits and Equipment Operating at Less than 50 volts; **Article 725**—Class 1, Class 2, and Class 3 Remote-Control, Signaling, and Power-Limited Circuits; **Article 727**—Instrumentation Tray Cable—Type ITC; **Article 760**—Fire Alarm Systems; **Article 770**—Optical Fiber Cables and Raceways; **Article 780**—Closed-Loop and Programmed Power Distribution.

CHAPTER 8
COMMUNICATIONS SYSTEMS

Article 800—Communications Circuits: **Article 810**—Radio and Television Equipment: **Article 820**—Community Antenna Television and Radio Distribution Systems.

9

TABLES AND EXAMPLES

A. TABLES

Notes to Tables

TABLE 1 Percent of Cross Section of Conduit and Tubing for Conductors

Number of Conductors	1	2	Over 2
All Conductor Types	53	31	40

Note 1: See Appendix C for the maximum number of conductors and fixture wires, all of the same size (total cross-sectional area including insulation), permitted in trade sizes of the applicable conduit or tubing.

Note 2: Table 1 applies only to complete conduit or tubing systems and is not intended to apply to sections of conduit or tubing used to protect exposed wiring from physical damage.

Note 3: Equipment grounding or bonding conductors, where installed, shall be included when calculating conduit or tubing fill. The actual dimensions of the equipment grounding or bonding conductor (insulated or bare) shall be used in the calculation.

Note 4: Where conduit or tubing nipples having a maximum length not to exceed 24 inches (610 mm) are installed between boxes, cabinets, and similar enclosures, the nipples shall be permitted to be filled to 60 percent of their total cross-sectional area, and Article 310, Note 8(a) of Notes to Ampacity Tables of 0 to 2000 Volts need not apply to this condition.

Note 5: For conductors not included in Chapter 9, such as multiconductor cables, the actual dimensions shall be used.

Note 6: For combinations of conductors of different sizes, use Tables 5 and 5A in Chapter 9 for dimensions of conductors and Table 4 for the applicable conduit or tubing dimensions.

Note 7: When calculating the maximum number of conductors permitted in a conduit or tubing, all of the same size (total cross-sectional area including insulation), the next higher whole number shall be used to determine the maximum number of conductors permitted when the calculation results in a decimal of 0.8 or larger.

Note 8: Where bare conductors are permitted by other sections of this Code, the dimensions for bare conductors in Table 8 of Chapter 9 shall be permitted.

Note 9: A multiconductor cable of two or more conductors shall be treated as a single conductor for calculating percentage conduit fill area. For cables that have elliptical cross sections, the cross-sectional area calculation shall be based on using the major diameter of the ellipse as a circle diameter.

(FPN): Table 1 is based on common conditions of proper cabling and alignment of conductors where the length of the pull and the number of bends are within reasonable limits. It should be recognized that, for certain conditions, a larger size conduit or a lesser conduit fill should be considered.

Note 10: When pulling three conductors or cables into a raceway, if the ratio of the raceway (inside diameter) to the conductor or cable (outside diameter) is between 2.8 and 3.2, jamming can occur and the next larger size raceway should be used. While jamming can occur when pulling four or more conductors or cables into a raceway, the probability is very low.

TABLE 4 Dimensions and Percent Area of Conduit and Tubing

(Areas of Conduit or Tubing for the Combinations of Wires Permitted in Table 1, Chapter 9)
Electrical Metallic Tubing

Trade Size in Inches	*Internal Diameter in Inches*	*Total Area 100% in Sq. In.*	*2 Wires 31% in Sq. In.*	*Over 2 Wires 40% in Sq. In.*	*1 Wire 53% in Sq. In.*
½	0.622	0.304	0.094	0.122	0.161
¾	0.824	0.533	0.165	0.213	0.283
1	1.049	0.864	0.268	0.346	0.458
1¼	1.380	1.496	0.464	0.598	0.793
1½	1.610	2.036	0.631	0.814	1.079
2	2.067	3.356	1.040	1.342	1.778
2½	2.731	5.858	1.816	2.343	3.105
3	3.356	8.846	2.742	3.538	4.688
3½	3.834	11.545	3.579	4.618	6.119
4	4.334	14.753	4.573	5.901	7.819

TABLE 4 (continued)

(Areas of Conduit or Tubing for the Combinations of Wires Permitted in Table 1, Chapter 9)
Electrical Nonmetallic Tubing

Trade Size in Inches	*Internal Diameter 100% in Sq. In.*	*Total Area 31% in Sq. In.*	*2 Wires 40% in Sq. In.*	*Over 2 Wires 53% in Sq. In.*	*1 Wire 53% in Sq. In.*
½	0.560	0.246	0.076	0.099	0.131
¾	0.760	0.454	0.141	0.181	0.240
1	1.000	0.785	0.243	0.314	0.416
1¼	1.340	1.410	0.437	0.564	0.747
1½	1.570	1.936	0.600	0.774	1.026
2	2.020	3.205	0.994	1.282	1.699
2½	—	—	—	—	—
3	—	—	—	—	—
3½	—	—	—	—	—
4	—	—	—	—	—

TABLE 4 (continued)

(Areas of Conduit or Tubing for the Combinations of Wires Permitted in Table 1, Chapter 9)
Flexible Metal Conduit

Trade Size in Inches	*Internal Diameter in Inches*	*Total Area 100% in Sq. In.*	*2 Wires 31% in Sq. In.*	*Over 2 Wires 40% in Sq. In.*	*1 Wire 53% in Sq. In.*
⅜	0.384	0.116	0.036	0.046	0.061
½	0.635	0.317	0.098	0.127	0.168
¾	0.824	0.533	0.165	0.213	0.282
1	1.020	0.817	0.253	0.327	0.433
1¼	1.275	1.277	0.396	0.511	0.677
1½	1.538	1.857	0.576	0.743	0.984
2	2.040	3.269	1.013	1.307	1.732
2½	2.500	4.909	1.522	1.964	2.602
3	3.000	7.069	2.191	2.827	3.746
3½	3.500	9.621	2.983	3.848	5.099
4	4.000	12.566	3.896	5.027	6.660

TABLE 4 (continued)

(Areas of Conduit or Tubing for the Combinations of Wires Permitted in Table 1, Chapter 9)
Intermediate Metal Conduit

Trade Size in Inches	*Internal Diameter in Inches*	*Total Area 100% in Sq. In.*	*2 Wires 31% in Sq. In.*	*Over 2 Wires 40% in Sq. In.*	*1 Wire 53% in Sq. In.*
⅜	—	—	—	—	—
½	0.660	0.342	0.106	0.137	0.181
¾	0.864	0.586	0.182	0.235	0.311
1	1.105	0.959	0.297	0.384	0.508
1¼	1.448	1.646	0.510	0.658	0.872
1½	1.683	2.223	0.689	0.889	1.178
2	2.150	3.629	1.125	1.452	1.923
2½	2.557	5.135	1.592	2.054	2.722
3	3.176	7.922	2.456	3.169	4.199
3½	3.671	10.584	3.281	4.234	5.610
4	4.166	13.631	4.226	5.452	7.224

TABLE 4 (continued)

(Areas of Conduit or Tubing for the Combinations of Wires Permitted in Table 1, Chapter 9)
Liquidtight Flexible Nonmetallic Conduit
(Type FNMC-B*)

Trade Size in Inches	*Internal Diameter in Inches*	*Total Area 100% in Sq. In.*	*2 Wires 31% in Sq. In.*	*Over 2 Wires 40% in Sq. In.*	*1 Wire 53% in Sq. In.*
⅜	0.494	0.192	0.059	0.077	0.102
½	0.632	0.314	0.097	0.125	0.166
¾	0.830	0.541	0.168	0.216	0.287
1	1.054	0.872	0.270	0.349	0.462
1¼	1.395	1.528	0.474	0.611	0.810
1½	1.588	1.979	0.614	0.792	1.049
2	2.033	3.245	1.006	1.298	1.720

* Corresponds to Section 351-22(2).

TABLE 4 (continued)

(Areas of Conduit or Tubing for the Combinations of Wires Permitted in Table 1, Chapter 9)
Liquidtight Flexible Nonmetallic Conduit
(Type FNMC-A*)

Trade Size in Inches	*Internal Diameter in Inches*	*Total Area 100% in Sq. In.*	*2 Wires 31% in Sq. In.*	*Over 2 Wires 40% in Sq. In.*	*1 Wire 53% in Sq. In.*
⅜	0.495	0.192	0.060	0.077	0.102
½	0.630	0.312	0.097	0.125	0.165
¾	0.825	0.535	0.166	0.214	0.283
1	1.043	0.854	0.265	0.341	0.452
1¼	1.383	1.501	0.465	0.600	0.796
1½	1.603	2.017	0.625	0.807	1.069
2	2.063	3.341	1.036	1.336	1.771

TABLE 4 (continued)

(Areas of Conduit or Tubing for the Combinations of Wires Permitted in Table 1, Chapter 9)
Liquidtight Flexible Metal Conduit

Trade Size in Inches	*Internal Diameter in Inches*	*Total Area 100% in Sq. In.*	*2 Wires 31% in Sq. In.*	*Over 2 Wires 40% in Sq. In.*	*1 Wire 53% in Sq. In.*
⅜	0.494	0.192	0.059	0.077	0.102
½	0.632	0.314	0.097	0.125	0.166
¾	0.830	0.541	0.168	0.216	0.287
1	1.054	0.872	0.270	0.349	0.462
1¼	1.395	1.528	0.474	0.611	0.810
1½	1.588	1.979	0.614	0.792	1.049
2	2.033	3.245	1.006	1.298	1.720
2½	2.493	4.879	1.513	1.952	2.586
3	3.085	7.475	2.317	2.990	3.962
3½	3.520	9.731	3.017	3.893	5.158
4	4.020	12.692	3.935	5.077	6.727
5	—	—	—	—	—
6	—	—	—	—	—

TABLE 4 (continued)

(Areas of Conduit or Tubing for the Combinations of Wires Permitted in Table 1, Chapter 9)
Rigid Metal Conduit

Trade Size in Inches	*Internal Diameter in Inches*	*Total Area 100% in Sq. In.*	*2 Wires 31% in Sq. In.*	*Over 2 Wires 40% in Sq. In.*	*1 Wire 53% in Sq. In.*
⅜	—	—	—	—	—
½	0.632	0.314	0.097	0.125	0.166
¾	0.836	0.549	0.170	0.220	0.291
1	1.063	0.888	0.275	0.355	0.470
1¼	1.394	1.526	0.473	0.610	0.309
1½	1.624	2.071	0.642	0.829	1.098
2	2.083	3.408	1.056	1.363	1.806
2½	2.489	4.866	1.508	1.946	2.579
3	3.090	7.499	2.325	3.000	3.975
3½	3.570	10.010	3.103	4.004	5.305
4	4.050	12.883	3.994	5.153	6.828
5	5.073	20.213	6.266	8.085	10.713
6	6.093	29.158	9.039	11.663	15.454

TABLE 4 (continued)

(Areas of Conduit or Tubing for the Combinations of Wires Permitted in Table 1, Chapter 9)
Rigid PVC Conduit, Schedule 80

Trade Size in Inches	*Internal Diameter in Inches*	*Total Area 100% in Sq. In.*	*2 Wires 31% in Sq. In.*	*Over 2 Wires 40% in Sq. In.*	*1 Wire 53% in Sq. In.*
½	0.526	0.217	0.067	0.087	0.115
¾	0.722	0.409	0.127	0.164	0.217
1	0.936	0.688	0.213	0.275	0.365
1¼	1.255	1.237	0.383	0.495	0.656
1½	1.476	1.711	0.530	0.684	0.907
2	1.913	2.874	0.891	1.150	1.1523
2½	2.290	4.119	1.277	1.647	2.183
3	2.864	6.442	1.997	2.577	3.414
3½	3.326	8.688	2.683	3.475	4.605
4	3.786	11.258	3.490	4.503	5.967
5	4.768	17.855	5.535	7.142	9.463
6	5.709	25.598	7.935	10.239	13.567

TABLE 4 (continued)

(Areas of Conduit or Tubing for the Combinations of Wires Permitted in Table 1, Chapter 9)
Rigid PVC Conduit Schedule 40 and HDPE Conduit

Trade Size in Inches	*Internal Diameter in Inches*	*Total Area 100% in Sq. In.*	*2 Wires 31% in Sq. In.*	*Over 2 Wires 40% in Sq. In.*	*1 Wire 53% in Sq. In.*
½	0.602	0.285	0.088	0.114	0.151
¾	0.804	0.508	0.157	0.203	0.269
1	1.029	0.832	0.258	0.333	0.441
1¼	1.360	1.453	0.450	0.581	0.770
1½	1.590	1.986	0.616	0.794	1.052
2	2.047	3.291	1.020	1.316	1.744
2½	2.445	4.695	1.455	1.878	2.488
3	3.042	7.268	2.253	2.907	3.852
3½	3.521	9.737	3.018	3.895	5.161
4	3.998	12.554	3.892	5.022	6.654
5	5.016	19.761	6.126	7.904	10.473
6	6.031	28.567	8.856	11.427	15.141

TABLE 4 (continued)

(Areas of Conduit or Tubing for the Combinations of Wires Permitted in Table 1, Chapter 9)
Type A, Rigid PVC Conduit

Trade Size in Inches	*Internal Diameter in Inches*	*Total Area 100% in Sq. In.*	*2 Wires 31% in Sq. In.*	*Over 2 Wires 40% in Sq. In.*	*1 Wire 53% in Sq. In.*
½	0.700	0.385	0.119	0.154	0.204
¾	0.910	0.650	0.202	0.260	0.345
1	1.175	1.084	0.336	0.434	0.575
1¼	1.500	1.767	0.548	0.707	0.937
1½	1.720	2.324	0.720	0.929	1.231
2	2.155	3.647	1.131	1.459	1.933
2½	2.635	5.453	1.690	2.181	2.890
3	3.230	8.194	2.540	3.278	4.343
3½	3.690	10.694	3.315	4.278	5.668
4	4.180	13.723	4.254	5.489	7.273
5	—	—	—	—	—
6	—	—	—	—	—

TABLE 4 (continued)

(Areas of Conduit or Tubing for the Combinations of Wires Permitted in Table 1, Chapter 9)
Type EB, PVC Conduit

Trade Size in Inches	*Internal Diameter in Inches*	*Total Area 100% in Sq. In.*	*2 Wires 31% in Sq. In.*	*Over 2 Wires 40% in Sq. In.*	*1 Wire 53% in Sq. In.*
½	—	—	—	—	—
¾	—	—	—	—	—
1	—	—	—	—	—
1¼	—	—	—	—	—
1½	—	—	—	—	—
2	2.221	3.874	1.201	1.550	2.053
2½	—	—	—	—	—
3	3.330	8.709	2.700	3.484	4.616
3½	3.804	11.365	3.523	4.546	6.024
4	4.289	14.448	4.479	5.779	7.657
5	5.316	22.195	6.881	8.878	11.764
6	6.336	31.530	9.774	12.612	16.711

TABLE 5 Dimensions of Insulated Conductors and Fixture Wire

Type: RFH–2, RHH, RHW, RHW–2, SF–1, SF–2, RH, FFH–2, SFF–1, SFF–2

Type	*Size*	*Approx. Diam. Inches*	*Approx. Area Sq. In.*
RFH–2	18	0.136	0.0145
FFH–2	16	0.148	0.0172
RH	14	0.163	0.0209
	12	0.182	0.026
RHW–2, RHH	14	0.193	0.0293
RHW	12	0.212	0.0353
RH, RHH	10	0.236	0.0437
RHW	8	0.326	0.0835
RHW–2	6	0.364	0.1041
	4	0.412	0.1333
	3	0.44	0.1521
	2	0.472	0.175
	1	0.582	0.266
	1/0	0.622	0.3039
	2/0	0.668	0.3505
	3/0	0.72	0.4072
	4/0	0.778	0.4754
	250	0.895	0.6291
	300	0.95	0.7988
	350	1.001	0.787
	400	1.048	0.8626
	500	1.133	1.0082
	600	1.243	1.2135
	700	1.314	1.3561
	750	1.348	1.4272
	800	1.38	1.4957
	900	1.444	1.6377
	1000	1.502	1.7719
	1250	1.729	2.3479
	1500	1.852	2.6938
	1750	1.966	3.0357
	2000	2.072	3.3719
SF–2	18	0.121	0.0115
SFF–2	16	0.133	0.0139
	14	0.148	0.0172
SF-1	18	0.091	0.0065
SFF–1			

TABLE 5 Dimensions of Insulated Conductors and Fixture Wire (Cont.)

Type: TF, THW, THW-2, TW, TFF, RFH–1, AF, THHW, RHW–2*, XF, XFF, RHH*, RHW*

Type	*Size*	*Approx. Diam. Inches*	*Approx. Area Sq. In.*
RFH–1, AF, XF, XFF	18	0.106	0.0088
AF, TF, TFF, XF, XFF	16	0.118	0.0109
TW, AF, XF, XFF	14	0.133	0.0139
TW	12	0.152	0.0181
	10	0.176	0.0243
	8	0.236	0.0437
RHH*, RHW*, RHW–2* THHW, THW, THW–2	14	0.163	0.0209
RHH*, RHW*, RHW–2*	12	0.182	0.026
THHW, THW, AF, XF, XFF	10	0.206	0.0333
RHH*, RHW*, RHW–2* THHW, THW, THW–2	8	0.266	0.0556
TW, THW	6	0.304	0.0726
THHW	4	0.352	0.0973
THW–2	3	0.38	0.1134
RHH*	2	0.412	0.1333
RHW*			
RHW–2*	1	0.492	0.1901
	1/0	0.532	0.2223
	2/0	0.578	0.2624
	3/0	0.63	0.3117
	4/0	0.688	0.3718
	250	0.765	0.4596
	300	0.82	0.5281
	350	0.871	0.5958
	400	0.918	0.6619
	500	1.003	0.7901
	600	1.113	0.9729
	700	1.184	1.101
	750	1.218	1.1652
	800	1.25	1.2272
	900	1.314	1.3561
	1000	1.372	1.4784
	1250	1.539	1.8602
	1500	1.662	2.1695
	1750	1.776	2.4773
	2000	1.882	2.7818

* Types RHH, RHW, & RHW–2 Without Outer Covering

TABLE 5 Dimensions of Insulated Conductors and Fixture Wire (Cont.)

Type: TFN, THHN, THWN, TFFN, THWN–2

Type	Size	Approx. Diam. Inches	Approx. Area Sq. In.
TFN	18	0.084	0.0055
TFFN	16	0.096	0.0133
THHN	14	0.111	0.0097
THWN	12	0.13	0.0133
THWN–2			
	10	0.164	0.0211
	8	0.216	0.0366
	6	0.254	0.0507
	4	0.324	0.0824
	3	0.352	0.0973
	2	0.384	0.1158
	1	0.446	0.1562
	1/0	0.486	0.1855
	2/0	0.532	0.2223
	3/0	0.584	0.2679
	4/0	0.642	0.3237
	250	0.711	0.397
	300	0.766	0.4608
	350	0.817	0.5242
	400	0.864	0.5863
	500	0.949	0.7073
	600	1.051	0.8676
	700	1.122	0.9887
	750	1.156	1.0496
	800	1.188	1.1085
	900	1.252	1.2311
	1000	1.31	1.3478

TABLE 5 Dimensions of Insulated Conductors and Fixture Wire (Cont.)

Type: FEP, FEPB, TFE, PF, PFA, PTFF, PGFF, PFF, PAF, PAFF, PFAH, PGF, PTF, Z, ZF, ZFF

Type	*Size*	*Approx. Diam. Inches*	*Approx. Area Sq. In.*
PF, PGFF			
PGF, PFF	18	0.086	0.0058
PTF, PAF	16	0.098	0.0075
PTFF, PAFF			
PF, PGFF	14	0.113	0.01
PGF, PFF			
PTF, PAF			
PTFF, PAFF			
TFE, FEP			
PFA, FEPB			
PFAH			
TFE, FEP	12	0.132	0.0137
PFA, FEPB	10	0.156	0.0191
PFAH			
	8	0.206	0.0333
	6	0.244	0.0468
	4	0.292	0.067
	3	0.32	0.0804
	2	0.352	0.0973
TFE	1	0.422	0.1399
PFAH			
TFE	1/0	0.462	0.1676
PFA	2/0	0.508	0.2027
PFAH, Z	3/0	0.56	0.2463
	4/0	0.618	0.3
ZF, ZFF	18	0.076	0.0045
	16	0.088	0.0061
Z, ZF, ZFF	14	0.103	0.0083
Z	12	0.122	0.0117
	10	0.156	0.0191
	8	0.196	0.0302
	6	0.234	0.043
	4	0.282	0.0625
	3	0.33	0.0855
	2	0.362	0.1029
	1	0.402	0.1269

TABLE 5 Dimensions of Insulated Conductors and Fixture Wire (Cont.)

Type: XHHW, ZW, XHH, XHHW–2, KF–1, KF–2, KFF–1, KFF–2

Type	*Size*	*Approx. Diam. Inches*	*Approx. Area Sq. In.*
XHHW, ZW	14	0.133	0.0139
XHHW–2	12	0.152	0.0181
XHH	10	0.176	0.0243
	8	0.236	0.0437
	6	0.274	0.059
	4	0.322	0.0814
	3	0.35	0.0962
	2	0.382	0.1146
XHHW	1	0.442	0.1534
XHHW–2	1/0	0.482	0.1825
XHH	2/0	0.528	0.219
	3/0	0.58	0.2642
	4/0	0.638	0.3197
	250	0.705	0.3904
	300	0.76	0.4536
	350	0.811	0.5166
	400	0.858	0.5782
	500	0.943	0.6984
	600	1.053	0.8709
	700	1.124	0.9923
	750	1.158	1.0532
	800	1.19	1.1122
	900	1.254	1.2351
	1000	1.312	1.3519
	1250	1.479	1.718
	1500	1.602	2.0157
	1750	1.716	2.3127
	2000	1.822	2.6073
KF–2	18	0.063	0.0031
KFF–2	16	0.075	0.0044
	14	0.09	0.0064
	12	0.109	0.0093
	10	0.133	0.0139
KF–1	18	0.057	0.0026
KFF–1	16	0.069	0.0037
	14	0.084	0.0055
	12	0.103	0.0083
	10	0.127	0.0127

TABLE 5A Compact Aluminum Building Wire Nominal Dimensions* and Areas

Bare Conductor			*Type THW and THHW*		*Type THHN*		*Type XHHW*		
Size AWG or kcmil	*Number of Standards*	*Diam. Inches*	*Approx. Diam. Inches*	*Approx. Area Sq. In.*	*Approx. Diam Inches*	*Approx. Area Sq. In.*	*Approx. Diam. Inches*	*Approx. Area Sq. In.*	*Size AWG or kcmil*
8	7	.134	.255	.0510	—	—	.224	.0394	8
6	7	.169	.290	.0660	.240	.0452	.260	.0530	6
4	7	.213	.335	.0881	.305	.0730	.305	.0730	4
2	7	.268	.390	.1194	.360	.1017	.360	.1017	2
1	19	.299	.465	.1698	.415	.1352	.415	.1352	1
1/0	19	.336	.500	.1963	.450	.1590	.450	.1590	1/0
2/0	19	.376	.545	.2332	.495	.1924	.490	.1885	2/0
3/0	19	.423	.590	.2733	.540	.2290	.540	.2290	3/0
4/0	19	.475	.645	.3267	.595	.2780	.590	.2733	4/0

TABLE 5A (Continued)

Bare Conductor			*Type THW and THHW*		*Type THHN*		*Type XHHW*		
Size AWG or kcmil	*Number of Standards*	*Diam. Inches*	*Approx. Diam. Inches*	*Approx. Area Sq. In.*	*Approx. Diam Inches*	*Approx. Area Sq. In.*	*Approx. Diam. Inches*	*Approx. Area Sq. In.*	*Size AWG or kcmil*
250	37	.520	.725	.4128	.670	.3525	.660	.3421	250
300	37	.570	.775	.4717	.720	.4071	.715	.4015	300
350	37	.616	.820	.5281	.770	.4656	.760	.4536	350
400	37	.659	.865	.5876	.815	.5216	.800	.5026	400
500	37	.736	.940	.6939	.885	.6151	.880	.6082	500
600	61	.813	1.050	.8659	.985	.7620	.980	.7542	600
700	61	.877	1.110	.9676	1.050	.8659	1.050	.8659	700
750	61	.908	1.150	1.0386	1.075	.9076	1.090	.9331	750
1000	61	1.060	1.285	1.2968	1.255	1.2370	1.230	1.1882	1000

* Dimensions are from industry sources

TABLE 8 Conductor Properties

Size AWG/kcmil	Area Cir. Mils	Conductors				DC Resistance at 75°C (167°F)		
		Stranding		Overall		Copper		Aluminum
		Quantity	Diam. In.	Diam. In.	Area In.^{2}	Uncoated ohm/kFt	Coated ohm/kFT	ohm/kFT
18	1620	1	—	0.040	0.001	7.77	8.08	12.8
18	1620	7	0.015	0.046	0.002	7.95	8.45	13.1
16	2580	1	—	0.051	0.002	4.89	5.08	8.05
16	2580	7	0.019	0.058	0.003	4.99	5.29	8.21
14	4110	1	—	0.064	0.003	3.07	3.19	5.06
14	4110	7	0.024	0.073	0.004	3.14	3.26	5.17
12	6530	1	—	0.081	0.005	1.93	2.01	3.18
12	6530	7	0.030	0.092	0.006	1.98	2.05	3.25
10	10380	1	—	0.102	0.008	1.21	1.26	2.00
10	10380	7	0.038	0.116	0.011	1.24	1.29	2.04

TABLE 8 (Continued)

		Conductors				DC Resistance at 75°C (167°F)		
		Stranding		Overall		Copper		Aluminum
Size AWG/kcmil	Area Cir. Mils	Quantity	Diam. In.	Diam. In.	Area $In.^2$	Uncoated ohm/kFt	Coated ohm/kFT	ohm/kFT
8	16510	1	—	0.128	0.013	0.764	0.786	1.26
8	16510	7	0.049	0.146	0.017	0.778	0.809	1.28
6	26240	7	0.061	0.184	0.027	0.491	0.510	0.808
4	41740	7	0.077	0.232	0.042	0.308	0.321	0.508
3	52620	7	0.087	0.260	0.053	0.245	0.254	0.403
2	66360	7	0.097	0.292	0.067	0.194	0.201	0.319
1	83690	19	0.066	0.332	0.087	0.154	0.160	0.253
1/0	105600	19	0.074	0.373	0.109	0.122	0.127	0.201
2/0	133100	19	0.084	0.418	0.137	0.0967	0.101	0.159
3/0	167800	19	0.094	0.470	0.173	0.0766	0.0797	0.126
4/0	211600	19	0.106	0.528	0.219	0.0608	0.0626	0.100

TABLE 8 (Continued)

		Conductors				DC Resistance at 75°C (167°F)		
		Stranding		Overall		Copper		Aluminum
Size AWG/kcmil	Area Cir. Mils	Quantity	Diam. In.	Diam. In.	Area $In.^2$	Uncoated ohm/kFt	Coated ohm/kFT	ohm/kFT
250	—	37	0.082	0.575	0.260	0.0515	0.0535	0.0847
300	—	37	0.090	0.630	0.312	0.0429	0.0446	0.0707
350	—	37	0.097	0.681	0.364	0.0367	0.0382	0.0605
400	—	37	0.104	0.728	0.416	0.0321	0.0331	0.0529
500	—	37	0.116	0.813	0.519	0.0258	0.0265	0.0424
600	—	61	0.099	0.893	0.626	0.0214	0.0223	0.0353
700	—	61	0.107	0.964	0.730	0.0184	0.0189	0.0303
750	—	61	0.111	0.998	0.782	0.0171	0.0176	0.0282
800	—	61	0.114	1.030	0.834	0.0161	0.0166	0.0265
900	—	61	0.122	1.094	0.940	0.0143	0.0147	0.0235
1000	—	61	0.128	1.152	1.042	0.0129	0.0132	0.0212
1250	—	91	0.117	1.289	1.305	0.0103	0.0106	0.0169

TABLE 8 (Continued)

		Conductors				*DC Resistance at 75°C (167°F)*		
		Stranding		*Overall*		*Copper*		*Aluminum*
Size AWG/kcmil	*Area Cir. Mils*	*Quantity*	*Diam. In.*	*Diam. In.*	*Area In.²*	*Uncoated ohm/ kFt*	*Coated ohm/ kFT*	*ohm/ kFT*
1500	—	91	0.128	1.412	1.566	0.00858	0.00883	0.0141
1750	—	127	0.117	1.526	1.829	0.00735	0.00756	0.0121
2000	—	127	0.126	1.632	2.092	0.00643	0.00662	0.0106

These resistance values are valid ONLY for the parameters as given. Using conductors having coated strands, different stranding type, and especially, other temperatures, change the resistance.

Formula for temperature change: $R_2 = R_1\ [1 + \alpha(T_2 - 75)]$ where $\alpha_{cu} = 0.00393$, $\alpha_{AI} = 0.00330$.

Conductors with compact and compressed stranding have about 9 percent and 3 percent, respectively, smaller bare conductor diameters than those shown. See Table 5A for actual compact cable dimensions.

The IACS conductives used: bare copper = 100%, aluminum = 61%.

Class B stranding is listed as well as solid for some sizes. Its overall diameter and area is that of its circumscribing circle.

(FPN) The construction information is per NEMA WC8-1988. The resistance is calculated per National Bureau of Standards Handbook 100, dated 1966, and Handbook 109, dated 1972.)

TABLE 9 AC Resistance and Reactance 600 V Cables, 3 Phase, 60 Hz, 75°C (167°F)—Three Single Conductors in Conduit

Size AWG/kcmil	Ohms to Neutral per 1000 feet																Size AWG/kcmil
	X_L (Reactance) for All Wires		AC Resistance for Uncoated Copper Wires			AC Resistance for Aluminum Wires			Effective Z at .85 PF for Uncoated Copper Wires			Effective Z at .85 PF for Aluminum Wires					
	PVC, Al. Conduit	Steel Conduit	PVC Conduit	Al. Conduit	Steel Conduit	PVC Conduit	Al. Conduit	Steel Conduit	PVC Conduit	Al. Conduit	Steel Conduit	PVC Conduit	Al. Conduit	Steel Conduit			
14	.058	.073	3.1	3.1	3.1	—	—	—	2.7	2.7	2.7	—	—	—			14
12	.054	.068	2.0	2.0	2.0	3.2	3.2	3.2	1.7	1.7	1.7	2.8	2.8	2.8			12
10	.050	.063	1.2	1.2	1.2	2.0	2.0	2.0	1.1	1.1	1.1	1.8	1.8	1.8			10
8	.052	.065	0.78	0.78	0.78	1.3	1.3	1.3	0.69	0.69	0.70	1.1	1.1	1.1			8
6	.051	.064	0.49	0.49	0.49	0.81	0.81	0.81	0.44	0.45	0.45	0.71	0.72	0.72			6
4	.048	.060	0.31	0.31	0.31	0.51	0.51	0.51	0.29	0.29	0.30	0.46	0.46	0.46			4
3	.047	.059	0.25	0.25	0.25	0.40	0.41	0.40	0.23	0.24	0.24	0.37	0.37	0.37			3
2	.045	.057	0.19	0.20	0.20	0.32	0.32	0.32	0.19	0.19	0.20	0.30	0.30	0.30			2
1	.046	.057	0.15	0.16	0.16	0.25	0.26	0.25	0.16	0.16	0.16	0.24	0.24	0.25			1
1/0	.044	.055	0.12	0.13	0.12	0.20	0.21	0.20	0.13	0.13	0.13	0.19	0.20	0.20			1/0
2/0	.043	.054	0.10	0.10	0.10	0.16	0.16	0.16	0.11	0.11	0.11	0.16	0.16	0.16			2/0
3/0	.042	.052	0.077	0.082	0.079	0.13	0.13	0.13	0.088	0.092	0.094	0.13	0.13	0.14			3/0
4/0	.041	.051	0.062	0.067	0.063	0.10	0.11	0.10	0.074	0.078	0.080	0.11	0.11	0.11			4/0

TABLE 9 (Continued)

Size AWG/ kcmil	X_L (Reactance) for All Wires		AC Resistance for Uncoated Copper Wires			AC Resistance for Aluminum Wires			Effective Z at .85 PF for Uncoated Copper Wires			Effective Z at .85 PF for Aluminum Wires			Size AWG/ kcmil
	Ohms to Neutral per 1000 feet														
	PVC, Al. Conduit	Steel Conduit	PVC Conduit	Al. Conduit	Steel Conduit	PVC Conduit	Al. Conduit	Steel Conduit	PVC Conduit	Al. Conduit	Steel Conduit	PVC Conduit	Al. Conduit	Steel Conduit	
250	.041	.052	0.052	0.057	0.054	0.085	0.090	0.086	0.066	0.070	0.073	0.094	0.098	0.10	250
300	.041	.051	0.044	0.049	0.045	0.071	0.076	0.072	0.059	0.063	0.065	0.082	0.086	0.088	300
350	.040	.050	0.038	0.043	0.039	0.061	0.066	0.063	0.053	0.058	0.060	0.073	0.077	0.080	350
400	.040	.049	0.033	0.038	0.035	0.054	0.059	0.055	0.049	0.053	0.056	0.066	0.071	0.073	400
500	.039	.048	0.027	0.032	0.029	0.043	0.048	0.045	0.043	0.048	0.050	0.057	0.061	0.064	500
600	.039	.048	0.023	0.028	0.025	0.036	0.041	0.038	0.040	0.044	0.047	0.051	0.055	0.058	600
750	.038	.048	0.019	0.024	0.021	0.029	0.034	0.031	0.036	0.040	0.043	0.045	0.049	0.052	750
1000	.037	.046	0.015	0.019	0.018	0.023	0.027	0.025	0.032	0.036	0.040	0.039	0.042	0.046	1000

Notes:

1. These values are based on the following constants: UL-type RHH wires with Class B stranding, in cradled configuration. Wire conductivities are 100% IACS copper, and 61% IACS aluminum, and aluminum conduit is 45% IACS. Capacitive reactance is ignored since it is negligible at these voltages.

These resistance values are valid only at 75°C (167°F) and for the parameters as given, but are representative for 600-volt wire types operating at 60 Hz.

2. "Effective Z" is defined as R cost(theta) + X sin(theta), where "theta" is the power factor angle of the circuit. Multiplying current by effective impedance gives a good approximation for line to neutral voltage drop. Effective impedance values in this table are valid only at .85 power factor.

For another circuit power factor (PF), effective impedance (Ze) can be calculated from R and X_L values given in this table as follows:

$$Ze = R \times PF + X_L \sin[\arccos(PF)]$$

TABLE 10 Expansion Characteristics of PVC Rigid Nonmetallic Conduit Coefficient of Thermal Expansion = 3.38×10^{-5}in./in./°F

Temperature Changes in Degrees F	*Length Change in Inches per 100 ft of PVC Conduit*	*Length Change Temperature Change in Degrees F*	*Length Change in Inches per 100 ft of PVC Conduit*	*Temperature Change in Degrees F*	*Length Change in Inches per 100 ft of PVC Conduit*	*Temperature Change in Degrees F*	*Length Change in Inches per 100 ft of PVC Conduit*
5	0.2	55	2.2	105	4.2	155	6.3
10	0.4	60	2.4	110	4.5	160	6.5
15	0.6	65	2.6	115	4.7	165	6.7
20	0.8	70	2.8	120	4.9	170	6.9
25	1.0	75	3.0	125	5.1	175	7.1
30	1.2	80	3.2	130	5.3	180	7.3
35	1.4	85	3.4	135	5.5	185	7.5
40	1.6	90	3.6	140	5.7	190	7.7
45	1.8	95	3.8	145	5.9	195	7.9
50	2.0	100	4.1	150	6.1	200	8.1

TABLES 11(A) AND 11(B)

For listing purposes, Tables 11(a) and 11(b) provide the required power source limitations for Class 2 and Class 3 power sources. Table 11(a) applies for alternating current sources, and Table 11(b) applies for direct current sources.

The power for Class 2 and Class 3 circuits shall be either (1) inherently limited requiring no overcurrent protection, or (2) not inherently limited requiring a combination of power source and overcurrent protection. Power sources designed for interconnnection shall be listed for the purpose.

As part of the listing, the Class 2 or Class 3 power source shall be durably marked where plainly visible to indicate the class of supply and its electrical rating. A Class 2 power source not suitable for wet location use shall be so marked. Exception: limited power circuits used by listed information technology equipment.

Overcurrent devices, where required, shall be located at the point where the conductor to be protected receives its supply and shall not be interchangeable with devices of higher ratings. The overcurrent device shall be permitted as an integral part of the power source.

NOTES FOR TABLES 11(A) AND 11(B)

Note 1. V_{max}, I_{max}, and VA_{max} are determined with the current limiting impedance in the circuit (not bypassed) as follows:

V_{max}: Maximum output voltage regardless of load with rated input applied.

I_{max}: Maximum output current under any noncapacitive load, including short circuit, and with overcurrent protection bypassed if used. Where a transformer limits the output current, I_{max} limits apply after one minute of operation. Where a current-limiting impedance, listed for the purpose, or as part of a listed product, is used in combination with a nonpower-limited transformer or a stored energy source, e.g., storage battery, to limit the output current, I_{max} limits apply after five seconds.

VA_{max}: Maximum volt-ampere output after one minute of operation regardless of load and overcurrent protection bypassed if used.

Note 2. For nonsinusoidal ac, V_{max} shall be not greater than 42.4 volts peak. Where wet contact (immersion not included) is likely to occur, Class 3 wiring methods shall be used or V_{max}

shall be not greater than: 15 volts for sinusoidal ac: 21.2 volts peak for nonsinusoidal ac.

Note 3. If the power source is a transformer, $(VA)_{max}$ is 350 or less when V_{max} is 15 or less.

Note 4. For dc interrupted at a rate of 10 to 200 Hz, V_{max} shall not be greater than 24.8 volts peak. Where wet contact (immersion not included) is likely to occur, Class 3 wiring methods shall be used or V_{max} shall not be greater than: 30 volts for continuous dc; 12.4 volts peak for dc that is interrupted at a rate of 10 to 200 Hz.

TABLES 12(A) AND 12(B)

For listing purposes, Tables 12(a) and 12(b) provide the required power source limitations for power-limited fire alarm sources. Table 12(a) applies for alternating current sources, and Table 12(b) applies for direct current sources.

The power for power-limited fire alarm circuits shall be either: (a) inherently limited requiring no overcurrent protection, or (b) not inherently limited requiring the power to be limited by a combination of power source and overcurrent protection.

As part of the listing, the PLFA power source shall be durably marked where plainly visible to indicate the class of supply and its electrical rating.

The overcurrent device, where required, shall be located at the point where the conductor to be protected receives its supply and shall not be interchangeable with devices of higher ratings. The overcurrent device shall be permitted as an integral part of the power source.

NOTES FOR TABLES 12(A) AND 12(B)

Note 1. V_{max}: Maximum output voltage regardless of load with rated input applied.

I_{max}: Maximum output current under any noncapacitive load, including short circuit, and with overcurrent protection bypassed if used. When a transformer limits the output current, I_{max} limits apply after one minute of operation. Where a current-limiting impedance, listed for the purpose, is used in combination with a nonpower-limited transformer or a stored energy source, e.g., storage battery, to limit the output current, I_{max} limits apply after five seconds.

TABLE 11(A) Class 2 and Class 3 Alternating Current Power Source Limitations

		Inherently Limited Power Source (Overcurrent protection not required)				*Not Inherently Limited Power Source (Overcurrent protection required)*			
Power Source		*Class 2*			*Class 3*	*Class 2*		*Class 3*	
Source voltage V_{max} (Volts) (Note 1)		0 through 20†	Over 20 and through 30†	Over 30 and through 150	Over 30 and through 100	0 through 20†	Over 20 and through 30†	Over 30 and through 100	Over 100 and through 150
Power Limitations $(VA)_{max}$ (Volt-Amps) (Note 1)		—	—	—	—	250 (Note 3)	250	250	N.A.
Current Limitations I_{max} (Amps) (Note 1)		8.0	8.0	0.005	$150/V_{max}$	$1000/V_{max}$	$1000/V_{max}$	$1000/V_{max}$	1.0
Maximum Overcurrent Protection (Amps)		—	—	—	—	5.0	$100/V_{max}$	$100/V_{max}$	1.0
Power Source Maximum Nameplate Rating	VA (Volt-Amps)	$5.0 \times V_{max}$	100	$.005 \times V_{max}$	100	$5.0 \times V_{max}$	100	100	100
	Current (Amps)	5.0	$100/V_{max}$	.005	$100/V_{max}$	5.0	$100/V_{max}$	$100/V_{max}$	$100/V_{max}$

† Voltage ranges shown are for sinusoidal ac in indoor locations or where wet contact is not likely to occur. For nonsinusoidal or wet contact conditions, see Note 2.

TABLE 11(B) Class 2 and Class 3 Direct Current Power Source Limitations

		Inherently Limited Power Source (Overcurrent protection not required)					Not Inherently Limited Power Source (Overcurrent protection required)			
Power Source		Class 2				Class 3	Class 2		Class 3	
Source voltage V_{max} (Volts) (Note 1)		0 through 20††	Over 20 and through 30††	Over 30 and through 60††	Over 60 and through 150	Over 60 and through 100	0 through 20††	Over 20 and through 60††	Over 60 and through 100	Over 100 and through 150
Power Limitations $(VA)_{max}$ (Volt-Amps) (Note 1)		—	—	—	—	—	250 (Note 3)	250	250	N.A.
Current Limitations $(VA)_{max}$ (Note 1)		8.0	8.0	$150/V_{max}$	0.005	$150/V_{max}$	$1000/V_{max}$	$1000/V_{max}$	$1000/V_{max}$	1.0
Maximum Overcurrent Protection (Amps)		—	—	—	—	—	5.0	$100/V_{max}$	$100/V_{max}$	1.0
Power Source Maximum Nameplate Rating	VA (Volt-Amps)	$5.0 \times V_{max}$	100	100	$.005 \times V_{max}$	100	$5.0 \times V_{max}$	100	100	100
	Current (Amps)	5.0	$100/V_{max}$	$100/V_{max}$	.005	$100/V_{max}$	5.0	$100/V_{max}$	$100/V_{max}$	$100/V_{max}$

††Voltage ranges shown are for continuous dc in indoor locations or where wet contact is not likely to occur. For interrupted or wet contact conditions, see Note 4.

TABLE 12(A) PLFA Alternating Current Power Source Limitations

Power Source		*Inherently Limited Power Source (Overcurrent protection not required)*			*Not Inherently Limited Power Source (Overcurrent protection required)*		
Circuit Voltage V_{max} (Volts) (Note 1)		0 through 20	Over 20 and through 30	Over 30 and through 100	0 through 20	Over 20 and through 100	Over 100 and through 150
Power Limitations $(VA)_{max}$ (Volt-Amps) (Note 1)		—	—	—	250 (Note 2)	250	N.A.
Current Limitations I_{max} (Amps) (Note 1)		8.0	8.0	$150/V_{max}$	$1000/V_{max}$	$1000/V_{max}$	1.0
Maximum Overcurrent Protection (Amps)		—	—	—	5.0	$100/V_{MAX}$	1.0
Power Source Maximum Nameplate Ratings	VA (Volt-Amps)	$5.0 \times V_{max}$	100	100	$5.0 \times V_{max}$	100	100
	Current (Amps)	5.0	$100/V_{max}$	$100/V_{max}$	5.0	$100/V_{max}$	$100/V_{max}$

Notes found on page 259.

TABLE 12(B) PLFA Direct Current Power Source Limitations

Power Source		*Inherently Limited Power Source (Overcurrent protection not required)*				*Not Inherently Limited Power Source (Overcurrent protection required)*		
Circuit Voltage V_{max} (Volts) (Note 1)		0 through 20	Over 20 and through 30	Over 30 and through 100	Over 100 and through 250	0 through 20	Over 20 and through 100	Over 100 and through 150
Power Limitations $(VA)_{max}$ (Volt-Amps) (Note 1)		—	—	—	—	250 (Note 2)	250	N.A.
Current Limitations I_{max} (Amps) (Note 1)		8.0	8.0	$150/V_{max}$	0.030	$1000/V_{max}$	$1000/V_{max}$	1.0
Maximum Overcurrent Protection (Amps)		—	—	—	—	5.0	$100/V_{max}$	1.0
Power Source Maximum Nameplate Ratings	VA (Volt-Amps)	$5.0 \times V_{max}$	100	100	$0.030 \times V_{max}$	$5.0 \times V_{max}$	100	100
	Current (Amps)	5.0	$100/V_{max}$	$100/V_{max}$	0.030	5.0	$100/V_{max}$	$100/V_{max}$

Notes found on page 259.

VA_{max}: Maximum volt-ampere output after one minute of operation regardless of load and overcurrent protection by-passed if used. Current limiting impedance shall not be by-passed when determining I_{max} and VA_{max}.

Note 2. If the power source is a transformer, $(VA)_{max}$ is 350 or less when V_{max} is 15 or less.

B. EXAMPLES

Selection of Conductors. In the following examples, the results are generally expressed in amperes. To select conductor sizes, refer to the 0 to 2000 Volt ampacity tables of Article 310, and the Notes that pertain to such tables.

Voltage. For uniform application of Articles 210, 215, and 220, a nominal voltage of 120, 120/240, 240, and 208Y/120 volts shall be used in computing the ampere load on the conductor.

Fractions of an Ampere. Except where the computations result in a major fraction of an ampere 0.5 or larger, such fractions shall be permitted to be dropped.

Power Factor. Calculations in the following examples are based, for convenience, on the assumption that all loads have the same power factor.

Ranges. For the computation of the range loads in these examples, Column A of Table 220-19 has been used. For optional methods, see Columns B and C of Table 220-19. Except where the computations result in a major fraction of a kilowatt (0.5 or larger), such fractions shall be permitted to be dropped.

SI Units. For SI units: one square foot = 0.093 square meter: one foot = 0.3048 meter.

Example No 1(a) One-Family Dwelling

The dwelling has a floor area of 1500 square feet exclusive of unoccupied cellar, unfinished attic, and open porches. Appliances are a 12-kW range and a 5.5-kW, 240-volt dryer. Assume range and dryer kW ratings equivalent to kVA ratings in accordance with Sections 220-18 and 220-19.

Compted Load [see Section 220-10(a)]

General lighting load: 1500 sq. ft. at 3 volt-amperes per sq. ft. = 4500 volt-amperes.

Minimum Number of Branch Circuits Required [see Section 220-4(b)]

General lighting load: 4500 volt-amperes ÷ 120 = 37.5. This requires three 15-A 2-wire or two 20-A 2-wire circuits.

Small appliance load: two 2-wire 20-A circuits [see Section 220-3(b)].

Laundry load: one 2-wire 20-A circuit [see Section 220-4(c)].

Minimum Size Feeder Required [see Section 220-10(a)]

General lighting	4500 VA
Small appliance load	3000 VA
Laundry	1500 VA
Total general light and small appliance	9000 VA
3000 volt-amperes at 100%	3000 VA
9000 – 3000 = 6000 volt-amperes at 35%	2100 VA
Net general lighting and small appliance load	5100 VA
Range load (see Table 220-19)	8000 VA
Dryer load (see Table 220-18)	5500 VA
Total load	18,600 VA

For 120/240-volt 3-wire single-phase service or feeder:

18,600 Va ÷ 240 V = 77.5 A

Net computed load exceeds 10 kVA. Service conductors shall be 100 amperes [see Section 230-42(b)(2)].

Neutral for Feeder and Service

Lighting and small appliance load	5100 VA
Range load, 8000 volt-amperes at 70%	5600 VA
Dryer load, 5500 volt-amperes at 70%	3850 VA
Total	14,550 VA

14,550 Va ÷ 240 V = 60.6 A

Example No. 1(b) One-Family Dwelling

Same conditions as Example No. 1(a), plus addition of one 6-ampere, 230-volt room air-conditioning unit and one 12-ampere, 115-volt room air-conditioning unit,* one 8-ampere, 115-volt rated disposal, and one 10-ampere, 120-volt rated dishwasher.* See Article 430 for general motors and Article

	Line A	*Neutral*	*Line B*
Amperes from Example No. 1(a)	78	61	78
One 230-V air conditioner	6		6
One 115-V air conditioner and 120 V dishwasher	12	12	10
One 115-V disposal		8	8
25% of largest motor (Section 430-24)	3	3	2
Amperes per line	99	84	104

* For feeder neutral, use largest of the two appliances for unbalance. Therefore, this dwelling shall be permitted to be served by a service rated 110 amperes.

440, Part G, for air-conditioning equipment. Motors have nameplate ratings of 115 V and 230 V for use on 120-V and 240-V nominal voltage systems.

From previous Example No. 1(a), feeder current is 78 amperes (3-wire 240 volts).

Example No. 2(a) Optional Calculation for One-Family Dwelling Heating Larger Than Air Conditioning [See Section 220-30 (a) and Table 220-30.]

Dwelling has a floor area of 1500 square feet exclusive of unoccupied cellar, unfinished attic, and open porches. It has a 12-kW range, a 2.5-kW water heater, a 1.2-kW dishwasher, 9 kW of electric space heating installed in five rooms, a 5-kW clothes dryer, and a 6-ampere, 230-volt room air-conditioning unit. Assume range, water heater, dishwasher, space heating, and clothes dryer kW ratings equivalent to kVA.

Air conditioner kVA is $6 \times 230 \div 1000 = 1.38$ kVA

1.38 kVA is less than the connected load of 9 kVA of space heating; therefore, the air-conditioner load need not be included in the service calculation (see Section 220-21).

1500 sq. ft. at 3 volt-amperes	4.5 kVA
Two 20-amp. appliance outlet circuits at 1500 volt-amperes each	3.0 kVA
Laundry circuit	1.5 kVA

Range (at nameplate rating)	12.0 kVA
Water heater	2.5 kVA
Dishwasher	1.2 kVA
Clothes dryer	5.0 kVA
	29.7 kVA

First 10 kVA of other load at 100% = 10.0 kVA
Remainder of other load at 40% (19.7 kVA × .4) = 7.88 kVA
Total of other load = 17.88 kVA
9 kVA of heat at 40% (9 × .4) = 3.6 kVA
Total load = 21.48 kVA

Calculated load for service size
21.48 kVA = 21,480 volt-amperes
21.480 VA ÷ 240 volts = 89.5 amperes

Therefore, this dwelling may be served by a 100-ampere service.

Feeder Neutral Load (per Section 220-22):

1500 sq. ft. at 3 volt-amperes	4500 VA
Three 20-amp. circuits at 1500 volt-amperes	4500 VA
Total	9000 VA
3000 volt-amperes at 100%	3000 VA
9000 VA – 3000 VA = 6000 volt-amperes at 35%	2100 VA
	5100 VA
Range—8 kVA at 70%	5600 VA
Clothes dryer—5 kVA at 70%	3500 VA
Dishwasher	1200 VA
Total	15,400 VA

15,400 VA ÷ 240 V = 64.2 A

Example No. 2(b) Optional Calculation for One-Family Dwelling Air Conditioning Larger Than Heating [See Section 220-30(a) and Table 220-30.]

Dwelling has a floor area of 1500 square feet exclusive of unoccupied cellar, unfinished attic, and open porches. It has two 20-ampere small appliance circuits, one 20-ampere laundry circuit, two 4-kW wall-mounted ovens, one 5.1-kW counter-mounted cooking unit, a 4.5-kW water heater, a 1.2-kW dishwasher, a 5-kW combination clothes washer and dryer, six 7-ampere, 230-volt room air-conditioning units, and a 1.5-kW permanently installed bathroom space heater. Assume wall-mounted ovens, counter-mounted cooking unit, water heater, dishwasher, and combination clothes washer and dryer kW ratings equivalent to kVA.

Air Conditioning kVA Calculation:

Total amperes: $6 \times 7 = 42.00$ A
$42 \times 240 \div 1000 = 10.08$ kVA of air-conditioned load; assume P.F. = 1.0.

Load Included at 100%:

Air conditioning (see below)
Space heater (omit: see Section 220-21)

Other Load:		kVA
1500 sq. ft. at 3 volt-amperes		4.5
Two 20-amp. small appliance circuits at 1500 volt-amperes		3.0
Laundry circuit		1.5
2 ovens		8.0
1 cooking unit		5.1
Water heater		4.5
Dishwasher		1.2
Washer/dryer		5.0
Total other load		32.8
1st 10 kVA at 100%		10.0 kVA
Remainder at 40% (22.8 kVA × .4)		9.12 kVA
Total other load		19.12 kVA
Other load	19.12 kVA	
Air conditioning	10.08 kVA	
	29.2 kVA = 29,200 volt-amperes	

29,200 VA ÷ 240 V = 122 A (service rating)

Feeder Neutral Load (per Section 220-22):

(It is assumed that the two-kVA wall-mounted ovens are supplied by one branch circuit, the 5.1-kVA counter-mounted cooking unit by a separate circuit.)

1500 sq. ft. at 3 volt-amperes	4500 VA
Three 20-amp circuits at 1500 volt-amperes	4500 VA
Total	9000 VA
3000 volt-amperes at 100%	3000 VA
9000 VA – 3000 VA = 6000 volt-amperes at 35%	2100 VA
Total	5100 VA

Two 4-kVA ovens plus one 5.1-kVA cooking unit totals 13.1 kVA. Table 220-19 permits 55% demand factor.

$$13.1 \text{ kVA} \times .55 = 7.2\text{-kVA feeder capacity}$$

7200 VA × 70% for neutral load	5040 VA
Clothes washer/dryer – 5 kVA × 70% for neutral load	3500 VA
Dishwasher	1200 VA
Total	14,840 VA

$$14{,}840 \text{ VA} \div 240 \text{ V} = 61.83, \text{ use } 62\text{A}$$

Example No. 2(c). Optional Calculation for One-Family Dwelling With Heat Pump. Single-Phase, 240/120-Volt Service. (See Section 220-30.)

Dwelling has a floor area of 2000 sq. ft. exclusive of unoccupied cellar, unfinished attic, and open porches. It has a 12-kW range, 4.5-kW water heater, a 1.2-kW dishwasher, a 5-kW clothes dryer, a 2½-ton (24-ampere) heat pump with 15-kW of back-up heat.

$$\text{Heat pump kVA is } 24 \times 240 \div 1000 = 5.76 \text{ kVA}$$

5.76 kVA is less than 15 kVA of the back up heat, therefore, the heat pump load need not be included in the service calculation. (See Table 220-30.)

2000 sq. ft. at 3 volt-amperes	6.0 kVA
Two 20-ampere appliance outlet circuits at 1500 volt-amperes each	3.0 kVA

Laundry circuit	1.5 kVA
Range (at nameplate rating)	12.0 kVA
Water heater	4.5 kVA
Dishwasher	1.2 kVA
Clothes dryer	5.0 kVA
	33.2 kVA
First 10 kVA of other load at 100% =	10.0 kVA
Remainder of other load at 40% (23.2 kVA × 0.4) =	9.28 kVA
Total of other load =	19.28 kVA

Heat pump and supplementary heat*

240 × 24 = 5760 VA

15-kW electric heat: 5760 VA + 15,000 VA = 20.760 VA or = 20.76 kVA

20.76 kVA at 65% = 13.49 kVA

* If supplementary heat is not on at the same time as heat pump, heat pump kVA need not be added to total.

Totals:

Other load	19.28 kVA
Heat pump and supplementary heat	13.49 kVA
Total	32.77 kVA

32.77 kVA ÷ 240 V = 136.5 A

Therefore, this dwelling unit may be served by a 150 ampere service.

Example No. 3. Store Building

A store 50 ft. by 60 ft., or 3000 sq. ft., has 30 ft. of show window. There are a total of 80 duplex receptacles. The service is 120/240-volt, single-phase (3-wire service). Actual connected lighting load, 8500 volt amperes.

Computed Load (Section 220-10):

Noncontinuous Loads:

Receptacle Load (Section (220-13)

80 receptacles at 180 VA =
14,400 VA

10,000 VA at 100%	10,000 volt-amperes
(14,400 – 10,000) VA at 50%	2200 volt-amperes
Total	12,200 volt-amperes

Continuous Loads:

General Lighting Load:	
3000 sq. ft. at 3 volt-amperes per sq. ft.	9000 volt-amperes
Show Window Lighting Load:	
30 ft. at 200 volt-amperes per foot	6000 volt-amperes
Outside sign circuit 1200 volt-amperes [Section 600-6(c)]	1200 volt-amperes
	16,200 volt-amperes
Total noncontinuous loads plus continuous loads	28,400 volt-amperes

Minimum Number of Branch Circuits Required:

General Lighting Load: Branch circuits need only to supply the actual connected load [Section 220-4(d)].

8500 VA × 1.25 = 10,625 volt-amperes
10,625 volt-amperes ÷ 240 volts =
44 amperes for 3-wire 120/240.

The lighting load shall be permitted to be served by 2-wire or 3-wire 15- or 20-ampere circuits with combined capacity equal to 44 amperes or greater for 3-wire circuits or 88 amperes or greater for 2-wire circuits. The feeder capacity as well as the number of branch-circuit positions available for lighting circuits in the panelboard must reflect the full calculated load of 9000 VA × 1.25 = 11,250 volt-amperes.

Show Window

6,000 VA × 1.25 = 7500 VA
7500 volt-amperes ÷ 240 volts
= 31 amperes for 3-wire, 120/240.

The show window lighting shall be permitted to be served by 2-wire or 3-wire circuits with a capacity equal to 31 amperes or greater for 3-wire circuits or 62 amperes or greater for 2-wire circuits.

Receptacles required by Section 210-62 are assumed to be included in the receptacle load above if these receptacles do not supply the show window lighting load.

Receptacles:

Receptacle Load:

14,400 volt-amperes ÷ 240 volts

= 60 amperes for 3-wire, 120/240.

The receptable load shall be permitted to be served by 2-wire or 3-wire circuits with a capacity equal to 60 amperes or greater for 3-wire circuits or 120 amperes or greater for 2-wire circuits.

Minimum Size Feeder (or Service) Overcurrent Protection [Section 220-10(b) or Section 239-90(a)]:

Total noncontinuous loads	12,200 volt-amperes
Total continuous load @ 125%	
16,200 × 1.25	20,250 volt-amperes
Total load	32,450 volt-amperes

32,450 VA ÷ 240 volts = 135 amperes

The next higher standard size is 150 amperes (Section 240-6).

Minimum Size Feeder (or Service Conductors) Required [Sections 215-3, 220-10(b), 230-42(a)]:

For 120/240-volt, 3-wire system: 32,450 volt-amperes ÷ 240 volts = 135 amperes. Service or feeder conductor is 1/0 Cu per Section 220-10(b) and Table 310-16 (75°C terminations).

In the example, 125 percent of the actual connected lighting load (8500 Va × 1.25 = 10,625 VA) is less than 125 percent of the load from Table 220-3(b), so the minimum lighting load from Table 220-3(b) is used in the calculation. Had the actual lighting load been greater than the value computed from Table 220-3(b), 125 percent of the actual connected lighting load would have been used.

Example No. 4(a) Multifamily Dwelling

Multifamily dwelling having 40 dwelling units.

Meters in two banks of 20 each and individual subfeeders to each dwelling unit.

One-half of the dwelling units are equipped with electric ranges not exceeding 12 kW each. Assume range kW rating equivalent to kVa rating in accordance with Section 220-19. Other half of ranges are gas ranges.

Area of each dwelling unit is 840 square feet.

Laundry facilities on premises available to all tenants. Add no circuit to individual dwelling unit. Add 1500 volt-amperes for each laundry circuit to house load and add to the example as a "house load."

Computed Load for Each Dwelling Unit (Article 220):

General lighting load: 840 sq. ft. at 3 volt-amperes per sq. ft.	2520 VA
Special appliance load: electric range (Section 220-19)	8000 VA

Minimum Number of Branch Circuits Required for Each Dwelling Unit (Section 220-4):

General lighting load: 2520 volt-amperes ÷ 120 = 21 amperes or two 15-ampere, 2-wire circuits; or two 20-ampere, 2-wire circuits.

Small appliance load: Two 2-wire circuits of No. 12 wire [see Section 220-4(b)].

Range circuit: 8000 volt-amperes ÷ 240 volts = 33 amperes or a circuit of two No. 8 and one No. 10 as permitted by Section 220-22. (See Section 210-19.)

Minimum Size Feeder Required for Each Dwelling Unit (Section 215-2):

Computed load (Article 220)	
General lighting load	2520 VA
Small appliance load, two 20-ampere circuits	3000 VA
Total computed load (without ranges)	5520 VA
Application of demand factor	
3000 volt-amperes at 100%	3000 VA
2520 volt-amperes at 35%	882 VA
Net computed load (without ranges)	3882 VA
Range load	8000 VA
Net computed load (with ranges)	11,882 VA

Size of Each Feeder (see Section 215-3)

For 120/240-volt, 3-wire system (without ranges):

Net computed load: 3882 VA ÷ 240 V = 16.2 A

For 120/240-volt, 3-wire system (with ranges):

Net computed load: 11,882 VA ÷ 240 V = 49.5 A

Feeder neutral:

Lighting and small appliance load	3882 VA
Range load, 8000 volt-amperes at 70% (see Section 220-22) (not included for apartments without electric range)	5600 VA
Net computed load (neutral)	9482 VA

9482 VA ÷ 240 V = 39.5 A

Minimum Size Feeders Required from Service Equipment to Meter Bank (for 20 Dwelling Units—10 with Ranges):

Total computed load:

Lighting and small appliance load, 20 × 5520 volt-amperes	110,400 VA
Application of demand factor:	
3000 volt-amperes at 100%	3,000 VA
107,400 volt-amperes at 35%	37,950 VA
Net computed lighting and small appliance load	40,590 VA
Range load, 10 ranges (less than 12 kVA, Col. A, Table 220-19)	25,000 VA
Net computed load (with ranges)	65,590 VA

For 120/240-volt, 3-wire system:

Net computed load: 65,590 VA ÷ 240 V = 273 A

Feeder neutral:

Lighting and small appliance load	40,590 VA
Range load, 25,000 volt-amperes at 70% (see Section 220-22)	17,500 VA
Computed load (neutral)	58,090 VA

58,090 VA ÷ 249 V = 242 A

Further demand factor (Section 220-22):

200 amperes at 100%	200 A
42 amperes at 70%	29 A
Net computed load (neutral)	229 A

Minimum Size Main Feeder (or Service Conductors) Required (Less Houseload) (for 40 Dwelling Units—20 with Ranges):

Total computed load:	
Lighting and small appliance load,	
40 × 5520 volt-amperes	220,800 VA
Application of demand factor:	
3000 volt-amperes at 100%	3,000 VA
117,000 volt-amperes at 35%	40,950 VA
100,800 volt-amperes at 25%	25,200 VA
Net computed lighting and small appliance load	69,150 VA
Range load, 20 ranges (less than 12 kVA, Col. A, Table 220-19)	35,000 VA
Net computed load	104,150 VA

For 120/240-volt, 3-wire system:

Net computed load: 104,150 VA ÷ 240 V = 434 A

Feeder neutral:	
Lighting and small appliance load	69,150 VA
Range load, 35,000 volt-amperes at 70% (see Section 220-22)	24,500 VA
Computed load (neutral)	93,650 VA

93,650 VA ÷ 240 V = 390 A

Further demand factor (see Section 220-22):	
200 amperes at 100%	200 A
190 amperes at 70%	133 A
Net computed load (neutral)	333 A

See Tables 310-16 through 310-31, Note 8 and 10.

Example No. 4(b) Optional Calculation for Multifamily Dwelling

Multifamily dwelling equipped with electric cooking and space heating or air conditioning and having 40 dwelling units.

Meters in two banks of 20 each plus house metering and individual subfeeders to each dwelling unit.

Each dwelling unit is equipped with an electric range of 8-kW nameplate rating, four 1.5-kW separately controlled 240-

volt electric space heaters, and a 2.5-kW 240-volt electric water heater. Assume range, space heater, and water heater kW ratings equivalent to kVA.

A common laundry facility is available to all tenants [Section 210-52(e), Exception No. 1].

Area of each dwelling unit is 840 square feet.

Computed Load for Each Dwelling Unit (Article 220):

General lighting load	
840 sq. ft. at 3 volt-amperes per sq. ft.	2520 VA
Electric range	8000 VA
Electric heat, 6 kVA (or air conditioning if larger)	6000 VA
Electric water heater	2500 VA

Minimum Number of Branch Circuits Required for Each Dwelling Unit:

General lighting load: 2520 volt-amperes ÷ 120 volts = 21 amperes or two 15-ampere 2-wire circuits or two 20-ampere 2-wire circuits.

Small appliance load: Two 2-wire circuits of No. 12 [see Section 220-4(b)].

Range circuit: 8000 volt-amperes × 80% ÷ 240 volts = 27 amperes on a circuit of three No. 10 conductors as permitted in Column C of Table 220-19.

Space heating: 6000 volt-amperes ÷ 240 volts = 25 amperes

No. of circuits (see Section 220-4).

Minimum Size Feeder Required for Each Dwelling Unit (Section 215-2):

Computed load (Article 220)	
General lighting load	2520 VA
Small appliance load, two 20-ampere circuits	3000 VA
Total computed load (without range and space heating)	5520 VA
Application of demand factor	
3000 volt-amperes at 100%	3000 VA
2520 volt-amperes at 35%	882 VA
Net computed load (without range and space heating)	3882 VA

Range load	6400 VA
Space heating (Section 220-15)	6000 VA
Water heater	2500 VA
Net computed load for individual dwelling unit	18,782 VA

For 120/240 volt, 3-wire system:

Net computed load: 18,782 VA ÷ 240 V = 78 A

Feeder neutral (Section 220-22)	
Lighting and small appliance load	3882 VA
Range load 6400 at 70% (see Section 220-22)	4480 VA
Space and water heating (no neutral) 240 volt	0 VA
Net computed load (neutral)	8362 VA

8362 VA ÷ 240 V = 35 A

Minimum Size Feeder Required from Service Equipment to Meter Bank for 20 Dwelling Units:

Total computed load:	
Lighting and small appliance load, 20 × 5520 volt-amperes	110,400 VA
Water and space heating load, 20 × 8500 volt-amperes	170,000 VA
Range load, 20 × 8000	160,000 VA
Net computed load (20 dwelling units)	440,400 VA
Net computed load using optional calculation (Table 220-32) 440,400 volt-amperes × .38	167,352 VA

167,352 VA ÷ 240 V = 697 A

Minimum Size Main Feeder Required (Less House Load) for 40 Dwelling Units:

Total computed load	
Lighting and small appliance load, 40 × 5520 volt-amperes	220,800 VA

Water and space heating, 40 × 8500 volt-amperes	340,000 VA
Range load, 40 × 8000 volt-amperes	320,000 VA
Net computed load (40 dwelling units)	880,800 VA
Net computed load using optional calculation (Table 220-32) 880,800 volt-amperes × .28	246,624 VA

246,624 VA ÷ 240 V = 1028 A

Feeder Neutral Load for Feeder from Service Equipment to Meter Bank (for 20 Dwelling Units):

Lighting and small appliance load, 20 × 5520 volt-amperes	110,400 VA
First 3000 volt-amperes at 100%	3,000 VA
107,400 volt-amperes at 35%	37,590 VA
Subtotal	40,590 VA
20 ranges = 35,000 volt-amperes at 70% (see Table 220-19 and Section 220-22)	24,500 VA
Total	65,090 VA

65,090 VA ÷ 240 V = 271 A

Further demand factor (Section 220-22)	
First 200 amperes at 100%	200 A
Balance: 71 amperes at 70%	50 A
Total	250 A

Feeder Neutral Load of Mains Feeder (Less House Load) for 40 Dwelling Units:

Lighting and small appliance load, 40 × 5520 volt-amperes	220,800 VA
First 3000 volt-amperes at 100%	3,000 VA
120,000 volt-amperes – 3000 volt-amperes = 117,000 volt-amperes at 35%	40,950 VA
220,800 volt-amperes – 120,000 volt-amperes = 100,800 volt-amperes at 25%	25,200 VA
Net computed lighting and small appliance load	69,150 VA

40 Ranges = 55,000 volt-amperes at 70%	38,500 VA
(See Table 220-19 and Section 220-22)	
Total	107,650 VA

107,650 VA ÷ 240 V = 449 A

Further demand factor (Section 220-22)	
First 200 amperes at 100%	200 A
Balance: 249 amperes at 70%	174 A
Total	374 A

Example No. 5(a) Multifamily Dwelling Served at 208^/120 Volts, Three Phase

All conditions and calculations the same as for Multifamily Dwelling [Example No. 4(a)] served 120/240 volts, single phase except as follows: service to each dwelling unit shall be two-phase legs and neutral.

Minimum Number of Branch Circuits Required for Each Dwelling Unit (Section 220-4):

Range circuit: 8000 volt-amperes ÷ 208 volts = 38 amperes or a circuit of two No. 8 and one No. 10 as permitted by Section 210-19(b).

Minimum Size Feeder Required for Each Dwelling Unit (Section 215-2):

For 120/208-volt, 3-wire system (without ranges)

Net Computed Load: 3882 volt-amperes ÷ 2 legs ÷ 120 volts/leg = 16.2 amperes

For 120-208-volt, 3-wire system (with ranges)

Net Computed Load: range; 8000 volt-amperes ÷ 208 volts = 38.5 amperes
Total load: range plus lighting = 54.7 amperes
Feeder neutral: range; 5600 volt-amperes ÷ 208 volts = 26.9 amperes
Total load: range plus lighting = 43.1 amperes

Minimum Size Feeders Required from Service Equipment to Meter Bank (for 20 Dwelling Units—10 with Ranges):

For 208Y/120-volt, 3-phase, 4-wire system:

Ranges: maximum number between any two phase legs = 4. Twice 4 = 8. Table 220-19 Demand = 23,000 volt-amperes.
Per phase demand: 23,000 volt-amperes ÷ 2 = 11,500 volt-amperes.
Equivalent 3-phase load = 34,500 volt-amperes.
Net computed load (total): 40,590 volt-amperes + 34,500 volt-amperes = 75,090 volt-amperes.

$$75{,}090 \text{ VA} \div (208)(1.732) = 208.4 \text{ A}$$

Feeder neutral size:

$$40{,}590 \text{ VA} + 34{,}500 \text{ VA at } 70\% = 64{,}700 \text{ VA}$$

Net computed neutral load:

$$64{,}700 \text{ VA} \div (208)(1.732) = 179.6 \text{ A}$$

Minimum Size Main Feeder (Less House Load) (for 40 Dwelling Units—20 with Ranges):

For 208Y/120-volt, 3-phase, 4-wire system:

Ranges: maximum number between any two phase legs = 7. Twice 7 = 14. Table 220-19 Demand = 29,000 volt-amperes.
Per phase demand: 29,000 ÷ 2 = 14,500 volt-amperes.
Equivalent 3-phase load: 43,500 volt-amperes.
Net computed load (total): 69,150 volt-amperes + 43,500 volt-amperes = 112,650 volt-amperes

$$112{,}650 \text{ VA} \div (208)(1.732) = 312.7 \text{ A}$$

Main feeder neutral size:

$$69{,}150 \text{ VA} + 43{,}500 \text{ VA at } 70\% = 99{,}600 \text{ VA}$$

99,600 volt-amperes ÷ (208)(1.732) = 276.5 amperes

Further demand factor (Section 220-22)

200 amperes at 100%	200.0 A
76.5 amperes at 70%	53.6 A
Net computed load	253.6 A

Example No. 5(b) Optional Calculation for Multifamily Dwelling Served at 208Y/120 Volts, Three Phase

All conditions and calculations the same as for Optional Calculations for Multifamily Dwelling [Example No. 4(b)] served at 120/40 volt, single phase except as follows:

Service to each dwelling unit shall be two-phase legs and neutral.

Minimum Number of Branch Circuits Required for Each Dwelling Unit (Section 220-4):

Range circuit: 8000 volt-amperes at 80% ÷ (208) = 30.7 amperes or a circuit of two No. 8 and one No. 10 as permitted by Section 220-22.

Space heating: 6000 volt-amperes ÷ 208 volts = 28.8 amperes. Two 20-ampere, 2-pole circuits required, No. 12.

Minimum Size Feeder Required for Each Dwelling Unit:

Computed Load (120/208-volt, 3-wire circuit)

Net Computed Load: 18,782 volt-amperes ÷ 208 volts	= 90.3 amperes
Net Computed Load: Lighting (line to neutral) 3882 volt-amperes ÷ 2 legs ÷ 120 volts/leg	16.2 amperes
Line to line: 14,900 volt-amperes ÷ 208 volts	71.6 amperes
Total load	87.8 amperes

Minimum Size Feeder Required for Service Equipment to Meter Bank (20 Dwelling Units)

Net computed load: 167,352 volt-amperes ÷ (208)(1.732) = 464.9 amperes.

Feeder neutral load:

65,090 VA ÷ (208)(1.732) = 180.65 A

Minimum Size Main Feeder Required (Less House Load) (40 Dwelling Units):

Net computed load: 246,624 volt-amperes ÷ (208)(1.732) = 684.6 amperes

Main feeder neutral load:

107,650 volt-amperes ÷ (208)(1.732) = 298.8 amperes

Further demand factor (Section 220-22)	
200 amperes at 100%	200.0 A
98.8 amperes at 70%	69.2 A
Net computed load	269.2 A

Example No. 6 Maximum Demand for Range Loads

Table 220-19, Column A, applies to ranges not over 12 kW. The application of Note 1 to ranges over 12 kW (and not over 27 kW) and Note 2 to ranges over 8 3/4 kW (and not over 27 kW) is illustrated in the following examples:

A. Ranges all the same rating (Table 220-19, Note 1)

Assume 24 ranges each rated 16 kW. Column A the maximum demand for 24 ranges of 12 kW rating is 39 kW.

16 kW exeeeds 12 kW by 4.
5% × 4 = 20% (5% increase for each kW in excess of 12).
39 kW × 20% = 7.8 kW increase.
39 + 7.8 = 46.8 kW: value to be used in selection of feeders.

B. Ranges of unequal rating

Assume:

5 ranges each rated 11 kW
2 ranges each rated 12 kW
20 ranges each rated 13.5 kW
3 ranges each rated 18 kW

5	×	12	=	60	Use 12 kW for range rated less than 12
2	×	12	=	24	
20	×	13.5	=	270	
3	×	18	=	54	
30				408 kW	

408 ÷ 30 = 13.6 kW (average to be used for computation)

From Column A the demand for 30 ranges of 12 kW rating is $15 + 30 = 45$ kW.

13.6 exceeds 12 by 1.6 (use 2).
$5\% \times 2 = 10\%$ (5% increase for each kW in excess of 12).
$45 \text{ kW} \times 10\% = 4.5$ kW increase.
$45 + 4.5 = 49.5$ kW: value to be used in selection of feeders.

Example No. 8 Motors, Conductors, Overload, and Short-Circuit and Ground-Fault Protection

(See Sections 240-6, 430-6, 430-7, 430-22, 430-23, 430-24, 430-32, 430-34, 430-52, and 430-62, and Tables 430-150 and 430-152.)

Determine the conductor size, the motor overload protection, the branch-circuit short-circuit and ground-fault protection, and the feeder protection, for one 25-horsepower, nameplate current 34.0 amperes, Design B, service factor 1.15 and two 30-horsepower wound-rotor induction motors (nameplate primary current 40 amperes, nameplate secondary current 65 amperes 40°C rise), on a 460-volt, 3-phase, 60-Hertz supply.

Conductor Loads

The full-load current value used to determine the ampacity of conductors for the 25-horsepower motor is 36 amperes [Section 430-6(a) and Table 430-150]. A full-load current of 34 amperes × 1.25 = 42.5 amperes (Section 430-22). The full-load current value used to determine the ampacity of primary conductors for each 30-horsepower motor is 40 amperes [Section 430-6(a) and Table 430-150]. A full-load primary current of 40 amperes × 1.25 = 50 amperes (Section 430-22). A full-load secondary current of 65 amperes × 1.25 = 81.25 amperes [Section 430-23(a)].

The feeder ampacity will be 125 percent of 40 plus 40 plus 34, or 124 amperes (Section 430-24).

Overload and Short-Circuit and Ground-Fault Protection

Overload. Where protected by a separate overload device, the 25-horsepower motor, with nameplate current of 34 amperes, shall have overload protection of not over 42.5 amperes [Sections 430-6(a) and 430-32(a)(1)]. Where protected by a separate overload device, the 30-horsepower motor, with nameplate current of 40 amperes, shall have overload protection of not over 50 amperes [Sections 430-6(a) and 430-32(a)(1)]. If the overload protection is not sufficient to start the motor or to carry the load, it shall be permitted to be increased according to Section 430-34. For a motor marked "thermally protected," overload protection is provided by the thermal protector [see Sections 430-7(a)(13) and 430-32(a)(2)].

Branch-Circuit Short-Circuit and Ground-Fault. The branch circuit of the 25-horsepower motor shall have branch-circuit short-circuit and ground-fault protection of not over 300 percent for a nontime-delay fuse (Table 430-152) or $3.00 \times 34 =$ 102 amperes. The next smaller standard size fuse is 110 amperes (Section 240-6 and 430-52(c), Exception No. 1). If a time-delay fuse is to be used because the motor will not start with a 110-ampere nontime-delay fuse, the maximum standard size permitted is 80 amperes [Section 430-52(c). Exception No. 2b]. If a nontime-delay fuse is to be used but the motor will not start with a 110-ampere nontime-delay fuse, the value is permitted to be increased to the next standard size, 125 amperes, because this rating does not exceed 400 percent [see Section 430-52(c), Exception No. 2a].

Feeder Circuit. The maximum rating of the feeder short-circuit and ground-fault protection device is based on the sum of the largest branch-circuit protective device (110 ampere fuse) plus the sum of the full-load currents of the other motors, or $110 + 40 + 40 = 190$ amperes. The nearest standard fuse that does not exceed the value is 175 amperes [Section 430-62(a)].

Example No. 9. Feeder Ampacity Determination for Generator Field Control

(See Sections 220-10(b); 430-24; 430-24 Exception No. 1; 620-13, 620-14, 620-61, Table 620-14 and Table 430-22(a) Exception 620-14.

Determine the conductor ampacity for a 460-volt 3-phase, 60 Hz ac feeder supplying a group of six elevators.

The 460-volt-ac drive motor nameplate rating of the largest MG set for one elevator is 40 horsepower, and 52 amperes and the remaining elevators, each have a 30-horsepower, 40-ampere, ac drive motor rating for their MG sets. In addition to a motor controller, each elevator has a separate motion/operation controller rated 10 amperes continuous to operate microprocessors, relays, power supplies, and the elevator car door operator. The MG sets are rated continuous.

Conductor ampacity is determined as follows:

(a) Per Sections 620-13(d) and 620-61(b)(1) use Table 430-22(a) Exception for intermittent duty (elevators). For intermittent duty using a continuous rated motor, the percentage of nameplate current rating to be used is 140 percent.

(b) For the 30-horsepower ac driven motors, 1.4 × 40 amps = 56 amperes. For the 40-horsepower ac drive motors, 1.4 × 52 amps = 73 amperes.

(c) The total conductor ampacity is the sum of all motor currents (73 amperes) + (5 × 56 amperes) = 353 amperes.

(d) Per Section 620-14 and Table 620-14, the conductor (feeder) ampacity shall be permitted to be reduced by the use of a demand factor. Constant loads are not included. (See Sections 620-14, FPN). For six elevators, the demand factor is .79. The feeder diverse ampacity is, therefore, .79 × 353 amperes = 279 amperes.

(e) Per Sections 430-24 and 220-10(b), the controller continuous current is 1.25 × 10 amperes = 12.5 amperes.

(f) The total feeder ampacity is the sum of the diverse current and all the controller continuous current. I total = 279 + (6 × 12.5) = 354 amperes.

(g) This ampacity shall be permitted to be used to select the wire size.

See Diagram for Example No. 9.

Example No. 10. Feeder Determination for Adjustable Speed Drive Control Ampacity [See Sections 220-10(b); 430-24; 430-24 Exception No. 1; 620-13, 620-14, 620-61 Tables 430-22(a), Exception and 620-14.]

Determine the conductor ampacity for a 460-volt, 3-phase, 60 Hz ac feeder supplying a group of six identical elevators. The

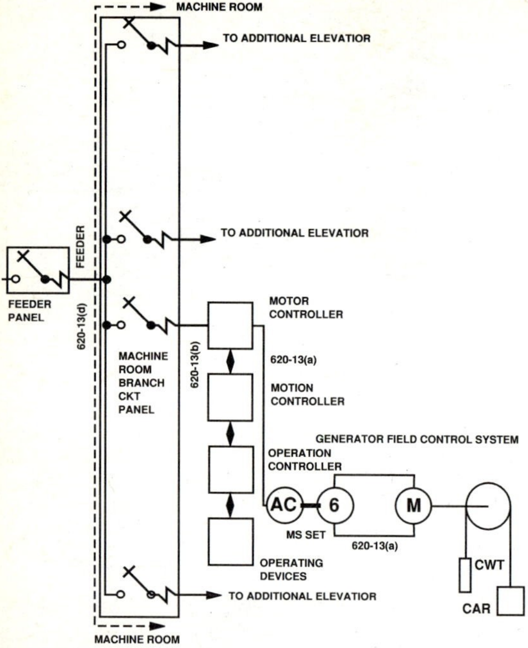
MACHINE ROOM
TO ADDITIONAL ELEVATIOR
FEEDER
TO ADDITIONAL ELEVATIOR
FEEDER
PANEL
620-13(d)
MOTOR
CONTROLLER
MACHINE
ROOM
BRANCH
CKT
PANEL
620-13(b)
620-13(a)
MOTION
CONTROLLER
GENERATOR FIELD CONTROL SYSTEM
OPERATION
CONTROLLER
AC
6
M
MS SET
620-13(a)
OPERATING
DEVICES
CWT
TO ADDITIONAL ELEVATIOR
CAR
MACHINE ROOM

Example 9

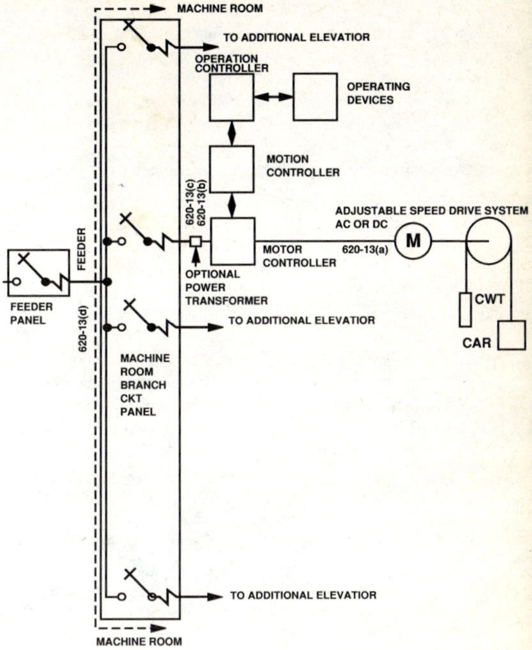
MACHINE ROOM
TO ADDITIONAL ELEVATIOR
OPERATION CONTROLLER
OPERATING DEVICES
MOTION CONTROLLER
620-13(c)
620-13(b)
ADJUSTABLE SPEED DRIVE SYSTEM AC OR DC
M
FEEDER
MOTOR CONTROLLER
620-13(a)
OPTIONAL POWER TRANSFORMER
FEEDER PANEL
CWT
620-13(d)
TO ADDITIONAL ELEVATIOR
CAR
MACHINE ROOM BRANCH CKT PANEL
TO ADDITIONAL ELEVATIOR
MACHINE ROOM

Example 10

system is adjustable speed SCR dc drive. The power transformers are external to the drive (motor controller) cabinet. Each elevator has a separate motion/operation controller connected to the load side of the main line disconnect switch rated 10 amperes (continuous) to operate microprocessors, relays, power supplies, and the elevator car door operator. Each transformer is rated 95 kVA with an efficiency of 90 percent.

Conductor ampacity is determined as follows:

(a) Calculate the nameplate rating of the transformer:

$$I = \frac{95 \times 1000}{\sqrt{3} \times 460 \times .90} = 133 \text{ amperes}$$

(b) Per Section 620-13(d), for six elevators, the total conductor ampacity is the sum of all the currents: 6 × 133 amperes = 798 amperes.

(c) Per Section 620-14 and Table 620-14, the conductor (feeder) ampacity shall be permitted to be reduced by the use of a demand factor. Constant loads are not included. (See FPN No. 2. 620-13). For six elevators, the demand factor is .79. The feeders diverse ampacity is, therefore, .79 × 798 amps = 630 amperes.

(d) Per Sections 430-24 and 220-10(b), the controller continuous current is 1.25 × 10 amperes = 12.5 amperes.

(e) The total feeder ampacity is the sum of the diverse current and the controller constant current. I total = 630 + (6 × 12) = 705 amperes.

(f) This ampacity shall be permitted to be used to select the wire size.

See Diagram for Example No. 10.

Appendix A

EXTRACTS

Appendix B

FORMULA APPLICATION, INFORMATION FOR AMPACITIES CALCULATED UNDER ENGINEERING SUPERVISION

Note: These appendices are listed for information. Refer to *NEC*® for full contents.

Appendix C

CONDUIT AND TUBING FILL TABLES FOR CONDUCTORS AND FIXTURE WIRE OF THE SAME SIZE

TABLE C1 Maximum Number of Conductors and Fixture Wires in Electrical Metallic Tubing
(Based on Table 1, Chapter 9)

Type Letters	*Conductor Size AWG/kcmil*	*Trade Sizes in Inches*									
		½	¾	*1*	*1¼*	*1½*	*2*	*2½*	*3*	*3½*	*4*
RH	14	6	10	16	28	39	64	112	169	221	282
	12	4	8	13	23	31	51	90	136	177	227
RHH, RHW, RHW-2	14	4	7	11	20	27	46	80	120	157	201
	12	3	6	9	17	23	38	66	100	131	167
RH, RHH, RHW, RHW-2	10	2	5	8	13	18	30	53	81	105	135
	8	1	2	4	7	9	16	28	42	55	70
	6	1	1	3	5	8	13	22	34	44	56
	4	1	1	2	4	6	10	17	26	34	44
	3	1	1	1	4	5	9	15	23	30	38
	2	1	1	1	3	4	7	13	20	26	33
	1	0	1	1	1	3	5	9	13	17	22

1/0	0	1	1	1	2	4	7	11	15	19
2/0	0	1	1	1	2	4	6	10	13	17
3/0	0	0	1	1	1	3	5	8	11	14
4/0	0	0	1	1	1	3	5	7	9	12
250	0	0	0	1	1	1	3	5	7	9
300	0	0	0	1	1	1	3	5	6	8
350	0	0	0	1	1	1	3	4	6	7
400	0	0	0	1	1	1	2	4	5	7
500	0	0	0	0	1	1	2	3	4	6
600	0	0	0	0	1	1	1	3	4	5
700	0	0	0	0	0	1	1	2	3	4
750	0	0	0	0	0	1	1	2	3	4
800	0	0	0	0	0	1	1	2	3	4
900	0	0	0	0	0	1	1	1	3	3
1000	0	0	0	0	0	1	1	1	2	3
1250	0	0	0	0	0	0	1	1	1	2
1500	0	0	0	0	0	0	1	1	1	1
1750	0	0	0	0	0	0	1	1	1	1
2000	0	0	0	0	0	0	1	1	1	1

TABLE C1 Maximum Number of Conductors and Fixture Wires in Electrical Metallic Tubing (continued)

Type Letters	*Conductor Size AWG/kcmil*	*Trade Sizes in Inches*									
		½	*¾*	*1*	*1¼*	*1½*	*2*	*2½*	*3*	*3½*	*4*
TW	14	8	15	25	43	58	96	168	254	332	424
	12	6	11	19	33	45	74	129	195	255	326
	10	5	8	14	24	33	55	96	145	190	243
	8	2	5	8	13	18	30	53	81	105	135
RHH*, RHW*, RHW-2*, THHW, THW, THW-2	14	6	10	16	28	39	64	112	169	221	282
RHH*, RHW*, RHW-2*, THHW, THW	12	4	8	13	23	31	51	90	136	177	227
	10	3	6	10	18	24	40	70	107	138	177
RHH*, RHW*, RHW-2*, THHW, THW, THW-2	8	1	4	6	10	14	24	42	63	83	106

RHH*, RHW*, RHW-2*, TW, THW, THHW, THW-2	6	1	3	4	8	11	18	32	48	63	81
	4	1	1	3	6	8	13	24	36	47	60
	3	1	1	3	5	7	12	20	31	40	52
	2	1	1	2	4	6	10	17	26	34	44
	1	1	1	1	3	4	7	12	18	24	31
	1/0	0	1	1	2	3	6	10	16	20	26
	2/0	0	1	1	1	3	5	9	13	17	22
	3/0	0	1	1	1	2	4	8	11	15	19
	4/0	0	0	1	1	1	3	6	9	12	16
	250	0	0	1	1	1	3	5	7	10	13
	300	0	0	1	1	1	2	4	6	8	11
	350	0	0	0	1	1	1	4	6	7	10
	400	0	0	0	1	1	1	3	5	7	9
	500	0	0	0	1	1	1	3	4	6	7
	600	0	0	0	1	1	1	2	3	4	6
	700	0	0	0	0	1	1	1	3	4	5
	750	0	0	0	0	1	1	1	3	4	5
	800	0	0	0	0	1	1	1	3	3	5
	900	0	0	0	0	0	1	1	2	3	4
	1000	0	0	0	0	0	1	1	2	3	4
	1250	0	0	0	0	0	1	1	1	2	3
	1500	0	0	0	0	0	1	1	1	1	2
	1750	0	0	0	0	0	0	1	1	1	2
	2000	0	0	0	0	0	0	1	1	1	1

* Types RHH, RHW, and RHW-2 without outer covering.

TABLE C1 Maximum Number of Conductors and Fixture Wires in Electrical Metallic Tubing (continued)

Type Letters	*Conductor Size AWG/kcmil*	*Trade Sizes in Inches*									
		½	*¾*	*1*	*1¼*	*1½*	*2*	*2½*	*3*	*3½*	*4*
THHN,	14	12	22	35	61	84	138	241	364	476	608
THWN,	12	9	16	26	45	61	101	176	266	347	443
THWN-2	10	5	10	16	28	38	63	111	167	219	279
	8	3	6	9	16	22	36	64	96	126	161
	6	2	4	7	12	16	26	46	69	91	116
	4	1	2	4	7	10	16	28	43	56	71
	3	1	1	3	6	8	13	24	36	47	60
	2	1	1	3	5	7	11	20	30	40	51
	1	1	1	1	4	5	8	15	22	29	37
	1/0	1	1	1	3	4	7	12	19	25	32
	2/0	0	1	1	2	3	6	10	16	20	26
	3/0	0	1	1	1	3	5	8	13	17	22
	4/0	0	1	1	1	2	4	7	11	14	18
	250	0	0	1	1	1	3	6	9	11	15
	300	0	0	1	1	1	3	5	7	10	13
	350	0	0	1	1	1	2	4	6	9	11
	400	0	0	0	1	1	1	4	6	8	10
	500	0	0	0	1	1	1	3	5	6	8

	600	0	0	0	1	1	1	2	4	5	7
	700	0	0	0	1	1	1	2	3	4	6
	750	0	0	0	0	1	1	1	3	4	5
	800	0	0	0	0	1	1	1	3	4	5
	900	0	0	0	0	1	1	1	3	3	4
	1000	0	0	0	0	1	1	1	2	3	4
FEP, FEPB, PFA, PFAH, TFE	14	12	21	34	60	81	134	234	354	462	590
	12	9	15	25	43	59	98	171	258	337	430
	10	6	11	18	31	42	70	122	185	241	309
	8	3	6	10	18	24	40	70	106	138	177
	6	2	4	7	12	17	28	50	75	98	126
	4	1	3	5	9	12	20	35	53	69	88
	3	1	2	4	7	10	16	29	44	57	73
	2	1	1	3	6	8	13	24	36	47	60
PFA, PFAH, TFE	1	1	1	2	4	6	9	16	25	33	42
PFA, PFAH, TFE, Z	1/0	1	1	1	3	5	8	14	21	27	35
	2/0	0	1	1	3	4	6	11	17	22	29
	3/0	0	1	1	2	3	5	9	14	18	24
	4/0	0	1	1	1	2	4	8	11	15	19

TABLE C1 Maximum Number of Conductors and Fixture Wires in Electrical Metallic Tubing (continued)

Type Letters	*Conductor Size AWG/kcmil*	*Trade Sizes in Inches*									
		½	*¾*	*1*	*1¼*	*1½*	*2*	*2½*	*3*	*3½*	*4*
Z	14	14	25	41	72	98	161	282	426	556	711
	12	10	18	29	51	69	114	200	302	394	504
	10	6	11	18	31	42	70	122	185	241	309
	8	4	7	11	20	27	44	77	117	153	195
	6	3	5	8	14	19	31	54	82	107	137
	4	1	3	5	9	13	21	37	56	74	94
	3	1	2	4	7	9	15	27	41	54	69
	2	1	1	3	6	8	13	22	34	45	57
	1	1	1	2	4	6	10	18	28	36	46
XHH, XHHW, XHHW-2, ZW	14	8	15	25	43	58	96	168	254	332	424
	12	6	11	19	33	45	74	129	195	255	326
	10	5	8	14	24	33	55	96	145	190	243
	8	2	5	8	13	18	30	53	81	105	135
	6	1	3	6	10	14	22	39	60	78	100
	4	1	2	4	7	10	16	28	43	56	72
	3	1	1	3	6	8	14	24	36	48	61
	2	1	1	3	5	7	11	20	31	40	51

XHH, XHHW, XHHW-2	1	1	1	1	4	5	8	15	23	30	38
	1/0	1	1	1	3	4	7	13	19	25	32
	2/0	0	1	1	2	3	6	10	16	21	27
	3/0	0	1	1	1	3	5	9	13	17	22
	4/0	0	1	1	1	2	4	7	11	14	18
	250	0	0	1	1	1	3	6	9	12	15
	300	0	0	1	1	1	3	5	8	10	13
	350	0	0	1	1	1	2	4	7	9	11
	400	0	0	0	1	1	1	4	6	8	10
	500	0	0	0	1	1	1	3	5	6	8
	600	0	0	0	1	1	1	2	4	5	6
	700	0	0	0	0	1	1	2	3	4	6
	750	0	0	0	0	1	1	1	3	4	5
	800	0	0	0	0	1	1	1	3	4	5
	900	0	0	0	0	1	1	1	3	3	4
	1000	0	0	0	0	0	1	1	2	3	4
	1250	0	0	0	0	0	1	1	1	2	3
	1500	0	0	0	0	0	1	1	1	1	3
	1750	0	0	0	0	0	0	1	1	1	2
	2000	0	0	0	0	0	0	1	1	1	1

TABLE C1 Maximum Number of Conductors and Fixture Wires in Electrical Metallic Tubing

Fixture Wires

Type Letters	*Conductor Size AWG/kcmil*	*Trade Sizes in Inches*					
		½	*¾*	*1*	*1½*	*1½*	*2*
FFH-2, RFH-2, RFHH-3	18 16	8 7	14 12	24 20	41 34	56 47	92 78
SF-2, SFF-2	18 16 14	10 8 7	18 15 12	30 25 20	52 43 34	71 58 47	116 96 78
SF-1, SFF-1	18	18	33	53	92	125	206
AF, RFH-1, RFHH-2, TF, TFF, XF, XFF	18	14	24	39	68	92	152
AF, RFHH-2, TF, TFF, XF, XFF	16	11	19	31	55	74	123
AF, XF, XFF	14	8	15	25	43	58	96
TFN, TFFN	18 16	22 17	38 29	63 48	108 83	148 113	244 186
PF, PFF, PGF, PGFF, PAF, PTF, PTFF, PAFF	18 16 14	21 16 12	36 28 21	59 46 34	103 79 60	140 108 81	231 179 134

TABLE C1 (continued)

Type Letters	Conductor Size AWG/kcmil	Trade Sizes in Inches					
		½	¾	1	1½	1½	2
ZF, ZFF,	18	27	47	77	133	181	298
ZHF, HF,	16	20	35	56	98	133	220
HFF	14	14	25	41	72	98	161
KF-2, KFF-2	18	39	69	111	193	262	433
	16	27	48	78	136	185	305
	14	19	33	54	93	127	209
	12	13	23	37	64	87	144
	10	8	15	25	43	58	96
KF-1, KFF-1	18	46	82	133	230	313	516
	16	33	57	93	161	220	362
	14	22	38	63	108	148	244
	12	14	25	41	72	98	161
	10	9	16	27	47	64	105
AX, XF, XFF	12	4	8	13	23	31	51
	10	3	6	10	18	24	40

Note: This table is for concentric stranded conductors only. For compact stranded conductors, Table C1A shall be used.

TABLE C1A Maximum Number of Compact Conductors in Electrical Metallic Tubing

(Based on Table 1, Chapter 9)

Type Letters	*Conductor Size AWG/kcmil*	*Conduit Trade Sizes in Inches*									
		½	*¾*	*1*	*1¼*	*1½*	*2*	*2½*	*3*	*3½*	*4*
THW,	8	2	4	6	11	16	26	46	69	90	115
THW-2,	6	1	3	5	9	12	20	35	53	70	89
THHW	4	1	2	4	6	9	15	26	40	52	67
	2	1	1	3	5	7	11	19	29	38	49
	1	1	1	1	3	4	8	13	21	27	34
	1/0	1	1	1	3	4	7	12	18	23	30
	2/0	0	1	1	2	3	5	10	15	20	25
	3/0	0	1	1	1	3	5	8	13	17	21
	4/0	0	1	1	1	2	4	7	11	14	18
	250	0	0	1	1	1	3	5	8	11	14
	300	0	0	1	1	1	3	5	7	9	12
	350	0	0	1	1	1	2	4	6	8	11
	400	0	0	0	1	1	1	4	6	8	10
	500	0	0	0	1	1	1	3	5	6	8
	600	0	0	0	1	1	1	2	4	5	7
	700	0	0	0	1	1	1	2	3	4	6
	750	0	0	0	0	1	1	1	3	4	5
	1000	0	0	0	0	1	1	1	2	3	4

THHN,	8	—	—	—	—	—	—	—	—	—	—
THWN,	6	2	4	7	13	18	29	52	78	102	130
THWN-2	4	1	3	4	8	11	18	32	48	63	81
	2	1	1	3	6	8	13	23	34	45	58
	1	1	1	2	4	6	10	17	26	34	43
	1/0	1	1	1	3	5	8	14	22	29	37
	2/0	1	1	1	3	4	7	12	18	24	30
	3/0	0	1	1	2	3	6	10	15	20	25
	4/0	0	1	1	1	3	5	8	12	16	21
	250	0	1	1	1	1	4	6	10	13	16
	300	0	0	1	1	1	3	5	8	11	14
	350	0	0	1	1	1	3	5	7	10	12
	400	0	0	1	1	1	2	4	6	9	11
	500	0	0	0	1	1	1	4	5	7	9
	600	0	0	0	1	1	1	3	4	6	7
	700	0	0	0	1	1	1	2	4	5	7
	750	0	0	0	1	1	1	2	4	5	6
	1000	0	0	0	0	1	1	1	3	3	4

TABLE C1A Maximum Number of Conductors and Fixture Wires in Electrical Metallic Tubing (continued)

Type Letters	*Conductor Size AWG/kcmil*	*Conduit Trade Sizes in Inches*									
		½	*¾*	*1*	*1¼*	*1½*	*2*	*2½*	*3*	*3½*	*4*
XHHW, XHHW-2	8	3	5	8	15	20	34	59	90	117	149
	6	1	4	6	11	15	25	44	66	87	111
	4	1	3	4	8	11	18	32	48	63	81
	2	1	1	3	6	8	13	23	34	45	58
	1	1	1	2	4	6	10	17	26	34	43
	1/0	1	1	1	3	5	8	14	22	29	37
	2/0	1	1	1	3	4	7	12	18	24	31
	3/0	0	1	1	2	3	6	10	15	20	25
	4/0	0	1	1	1	3	5	8	13	17	21
	250	0	1	1	1	2	4	7	10	13	17
	300	0	0	1	1	1	3	6	9	11	14
	350	0	0	1	1	1	3	5	8	10	13
	400	0	0	1	1	1	2	4	7	9	11
	500	0	0	0	1	1	1	4	6	7	9
	600	0	0	0	1	1	1	3	4	6	8
	700	0	0	0	1	1	1	2	4	5	7
	750	0	0	0	1	1	1	2	3	5	6
	1000	0	0	0	0	1	1	1	3	4	5

Definition: Compact stranding is the result of a manufacturing process where the standard conductor is compressed to the extent that the interstices (voids between strand wires) are virtually eliminated.

TABLE C2 Maximum Number of Conductors and Fixture Wires in Electrical Nonmetallic Tubing

(Based on Table 1, Chapter 9)

Type Letters	Conductor Size AWG/kcmil	Trade Sizes in Inches					
		½	¾	1	1¼	1½	2
RH	14	4	8	15	27	37	61
	12	3	7	12	21	29	49
RHH, RHW, RHW-2	14	3	6	10	19	26	43
	12	2	5	9	16	22	36
RH, RHH, RHW, RHW-2	10	1	4	7	13	17	29
	8	1	1	3	6	9	15
	6	1	1	3	5	7	12
	4	1	1	2	4	6	9
	3	1	1	1	3	5	8
	2	0	1	1	3	4	7
	1	0	1	1	1	3	5
	1/0	0	0	1	1	2	4
	2/0	0	0	1	1	1	3
	3/0	0	0	1	1	1	3
	4/0	0	0	1	1	1	2
	250	0	0	0	1	1	1
	300	0	0	0	1	1	1
	350	0	0	0	1	1	1
	400	0	0	0	1	1	1
	500	0	0	0	0	1	1
	600	0	0	0	0	1	1
	700	0	0	0	0	0	1
	750	0	0	0	0	0	1
	800	0	0	0	0	0	1
	900	0	0	0	0	0	1
	1000	0	0	0	0	0	1
	1250	0	0	0	0	0	0
	1500	0	0	0	0	0	0
	1750	0	0	0	0	0	0
	2000	0	0	0	0	0	0
TW	14	7	13	22	40	55	92
	12	5	10	17	31	42	71
	10	4	7	13	23	32	52
	8	1	4	7	13	17	29

TABLE C2 Maximum Number of Conductors and Fixture Wires in Electrical Nonmetallic Tubing (continued)

Type Letters	Conductor Size AWG/kcmil	Trade Sizes in Inches					
		½	¾	1	1¼	1½	2
RHH*, RHW*, RHW-2*, THHW, THW, THW-2	14	4	8	15	27	37	61
RHH*, RHW*, RHW-2*, THHW, THW	12	3	7	12	21	29	49
	10	3	5	9	17	23	38
RHH*, RHW*, RHW-2*, THHW, THW, THW-2	8	1	3	5	10	14	23
RHH*, RHW*, RHW-2*, TW, THW, THHW, THW-2	6	1	2	4	7	10	17
	4	1	1	3	5	8	13
	3	1	1	2	5	7	11
	2	1	1	2	4	6	9
	1	0	1	1	3	4	6
	1/0	0	1	1	2	3	5
	2/0	0	1	1	1	3	5
	3/0	0	0	1	1	2	4
	4/0	0	0	1	1	1	3
	250	0	0	1	1	1	2
	300	0	0	0	1	1	2
	350	0	0	0	1	1	1
	400	0	0	0	1	1	1
	500	0	0	0	1	1	1
	600	0	0	0	0	1	1
	700	0	0	0	0	1	1
	750	0	0	0	0	1	1
	800	0	0	0	0	1	1
	900	0	0	0	0	0	1
	1000	0	0	0	0	0	1
	1250	0	0	0	0	0	1
	1500	0	0	0	0	0	0
	1750	0	0	0	0	0	0
	2000	0	0	0	0	0	0

* Types RHH, RHW, and RHW-2 without outer covering.

TABLE C2 (continued)

Type Letters	Conductor Size AWG/kcmil	Trade Sizes in Inches					
		½	¾	1	1¼	1½	2
THHN, THWN, THWN-2	14	10	18	32	58	80	132
	12	7	13	23	42	58	96
	10	4	8	15	26	36	60
	8	2	5	8	15	21	35
	6	1	3	6	11	15	25
	4	1	1	4	7	9	15
	3	1	1	3	5	8	13
	2	1	1	2	5	6	11
	1	1	1	1	3	5	8
	1/0	0	1	1	3	4	7
	2/0	0	1	1	2	3	5
	3/0	0	1	1	1	3	4
	4/0	0	0	1	1	2	4
	250	0	0	1	1	1	3
	300	0	0	1	1	1	2
	350	0	0	0	1	1	2
	400	0	0	0	1	1	1
	500	0	0	0	1	1	1
	600	0	0	0	1	1	1
	700	0	0	0	0	1	1
	800	0	0	0	0	1	1
	900	0	0	0	0	1	1
	1000	0	0	0	0	0	1
FEP, FEPB, PFA, PFAH, TFE	14	10	18	31	56	77	128
	12	7	13	23	41	56	93
	10	5	9	16	29	40	67
	8	3	5	9	17	23	38
	6	1	4	6	12	16	27
	4	1	2	4	8	11	19
	3	1	1	4	7	9	16
	2	1	1	3	5	8	13
PFA, PFAH, TFE	1	1	1	1	4	5	9

TABLE C2 Maximum Number of Conductors and Fixture Wires in Electrical Nonmetallic Tubing (continued)

Type Letters	*Conductor Size AWG/kcmil*	*Trade Sizes in Inches*					
		½	*¾*	*1*	*1¼*	*1½*	*2*
PFA, PFAH, TFE, Z	1/0	0	1	1	3	4	7
	2/0	0	1	1	2	4	6
	3/0	0	1	1	1	3	5
	4/0	0	1	1	1	2	4
Z	14	12	22	38	68	93	154
	12	8	15	27	48	66	109
	10	5	9	16	29	40	67
	8	3	6	10	18	25	42
	6	1	4	7	13	18	30
	4	1	3	5	9	12	20
	3	1	1	3	6	9	15
	2	1	1	3	5	7	12
	1	1	1	2	4	6	10
XHH, XHHW, XHHW-2, ZW	14	7	13	22	40	55	92
	12	5	10	17	31	42	71
	10	4	7	13	23	32	52
	8	1	4	7	13	17	29
	6	1	3	5	9	13	21
	4	1	1	4	7	9	15
	3	1	1	3	6	8	13
	2	1	1	2	5	6	11

TABLE C2 (continued)

Type Letters	*Conductor Size AWG/kcmil*	*Trade Sizes in Inches*					
		½	¾	*1*	*1¼*	*1½*	*2*
XHH, XHHW, XHHW-2	1	1	1	1	3	5	8
	1/0	0	1	1	3	4	7
	2/0	0	1	1	2	3	6
	3/0	0	1	1	1	3	5
	4/0	0	0	1	1	2	4
	250	0	0	1	1	1	3
	300	0	0	1	1	1	3
	350	0	0	1	1	1	2
	400	0	0	0	1	1	1
	500	0	0	0	1	1	1
	600	0	0	0	1	1	1
	700	0	0	0	0	1	1
	750	0	0	0	0	1	1
	800	0	0	0	0	1	1
	900	0	0	0	0	1	1
	1000	0	0	0	0	0	1
	1250	0	0	0	0	0	1
	1500	0	0	0	0	0	1
	1750	0	0	0	0	0	0
	2000	0	0	0	0	0	0

TABLE C2 Maximum Number of Conductors and Fixture Wires in Electrical Nonmetallic Tubing

Fixture Wires

Type Letters	*Conductor Size AWG/kcmil*	*Trade Sizes in Inches*					
		½	*¾*	*1*	*1¼*	*1½*	*2*
FFH-2, RFH-2, RFHH-3	18 16	6 5	12 10	21 18	39 32	53 45	88 74
SF-2, SFF-2	18 16 14	8 7 5	15 13 10	27 22 18	49 40 32	67 55 45	111 92 74
SF-1, SFF-1	18	15	28	48	86	119	197
AF, RFH-1, RFHH-2, TF, TFF, XF, XFF	18	11	20	35	64	88	145
AF, RFHH-2, TF, TFF, XF, XFF	16	9	16	29	51	71	117
AF, XF, XFF	14	7	13	22	40	55	92
TFN, TFFN	18 16	18 13	33 25	57 43	102 78	141 107	233 178
PF, PFF, PGF, PGFF, PAF, PTF, PTFF, PAFF	18 16 14	17 13 10	31 24 18	54 42 31	97 75 56	133 103 77	221 171 128

TABLE C2 (continued)

Type Letters	Conductor Size AWG/kcmil	Trade Sizes in Inches					
		½	¾	1	1¼	1½	2
ZF, ZFF, ZHF, HF, HFF	18	22	40	70	125	172	285
	16	16	29	51	92	127	210
	14	12	22	38	68	93	154
KF-2, KFF-2	18	31	58	101	182	250	413
	16	22	41	71	128	176	291
	14	15	28	49	88	121	200
	12	10	19	33	60	83	138
	10	7	13	22	40	55	92
KF-1, KFF-1	18	38	69	121	217	298	493
	16	26	49	85	152	209	346
	14	18	33	57	102	141	233
	12	12	22	38	68	93	154
	10	7	14	24	44	61	101
AF, XF, XFF	12	3	7	12	21	29	49
	10	3	5	9	17	23	38

Note: This table is for concentric stranded conductors only. For compact stranded conductors, Table C2A shall be used.

TABLE C2A Maximum Number of Compact Conductors in Electrical Nonmetallic Tubing
(Based on Table 1, Chapter 9)

Type Letters	Conductor Size AWG/kcmil	Trade Sizes in Inches					
		½	¾	1	1¼	1½	2
THW, THW-2, THHW	8	1	3	6	11	15	25
	6	1	2	4	8	11	19
	4	1	1	3	6	8	14
	2	1	1	2	4	6	10
	1	0	1	1	3	4	7
	1/0	0	1	1	3	4	6
	2/0	0	1	1	2	3	5
	3/0	0	1	1	1	3	4
	4/0	0	0	1	1	2	4
	250	0	0	1	1	1	3
	300	0	0	1	1	1	2
	350	0	0	0	1	1	2
	400	0	0	0	1	1	1
	500	0	0	0	1	1	1
	600	0	0	0	1	1	1
	700	0	0	0	0	1	1
	750	0	0	0	0	1	1
	1000	0	0	0	0	0	1
THHN, THWN, THWN-2	8	-	-	-	-	-	-
	6	1	4	7	12	17	28
	4	1	2	4	7	10	17
	2	1	1	3	5	7	12
	1	1	1	2	4	5	9
	1/0	1	1	1	3	5	8
	2/0	0	1	1	3	4	6
	3/0	0	1	1	2	3	5
	4/0	0	1	1	1	2	4
	250	0	0	1	1	1	3
	300	0	0	1	1	1	3
	350	0	0	1	1	1	2
	400	0	0	0	1	1	2
	500	0	0	0	1	1	1
	600	0	0	0	1	1	1
	700	0	0	0	1	1	1
	750	0	0	0	1	1	1
	1000	0	0	0	0	1	1

TABLE C2A Maximum Number of Compact Conductors in Electrical Nonmetallic Tubing (continued)

Type Letters	Conductor Size AWG/kcmil	Trade Sizes in Inches					
		½	¾	1	1¼	1½	2
XHHW, XHHW-2	8	2	4	8	14	19	32
	6	1	3	6	10	14	24
	4	1	2	4	7	10	17
	2	1	1	3	5	7	12
	1	1	1	2	4	5	9
	1/0	1	1	1	3	5	8
	2/0	0	1	1	3	4	7
	3/0	0	1	1	2	3	5
	4/0	0	1	1	1	3	4
	250	0	0	1	1	1	3
	300	0	0	1	1	1	3
	350	0	0	1	1	1	3
	400	0	0	1	1	1	2
	500	0	0	0	1	1	1
	600	0	0	0	1	1	1
	700	0	0	0	1	1	1
	750	0	0	0	1	1	1
	1000	0	0	0	0	1	1

Defintion: Compact stranding is the result of a manufacturing process where the standard conductor is compressed to the extent that the interstices (voids between strand wires) are virtually eliminated.

TABLE C3 Maximum Number of Conductors and Fixture Wires in Flexible Metal Conduit
(Based on Table 1, Chapter 9)

Type Letters	*Conductor Size AWG/kcmil*	*Conduit Trade Sizes in Inches*									
		½	¾	1	1¼	1½	2	2½	3	3½	4
RH	14	6	10	15	24	35	62	94	135	184	240
	12	5	8	12	19	28	50	75	108	148	193
RHH, RHW, RHW-2	14	4	7	11	17	25	44	67	96	131	171
	12	3	6	9	14	21	37	55	80	109	142
RH, RHH, RHW, RHW-2	10	3	5	7	11	17	30	45	64	88	115
	8	1	2	4	6	9	15	23	34	46	60
	6	1	1	3	5	7	12	19	27	37	48
	4	1	1	2	4	5	10	14	21	29	37
	3	1	1	1	3	5	8	13	18	25	33
	2	1	1	1	3	4	7	11	16	22	28
	1	0	1	1	1	2	5	7	10	14	19
	1/0	0	1	1	1	2	4	6	9	12	16
	2/0	0	1	1	1	1	3	5	8	11	14
	3/0	0	0	1	1	1	3	5	7	9	12

	1/0	0	1	1	1	3	6	9	12	17	22
	2/0	0	1	1	1	3	5	7	10	14	19
	3/0	0	1	1	1	2	4	6	9	12	16
	4/0	0	0	1	1	1	3	5	7	10	13
	250	0	0	1	1	1	3	4	6	8	11
	300	0	0	1	1	1	2	3	5	7	9
	350	0	0	0	1	1	1	3	4	6	8
	400	0	0	0	1	1	1	3	4	6	7
	500	0	0	0	1	1	1	2	3	5	6
	600	0	0	0	0	1	1	1	3	4	5
	700	0	0	0	0	1	1	1	2	3	4
	750	0	0	0	0	1	1	1	2	3	4
	800	0	0	0	0	1	1	1	1	3	4
	900	0	0	0	0	0	1	1	1	3	3
	1000	0	0	0	0	0	1	1	1	2	3
	1250	0	0	0	0	0	1	1	1	1	2
	1500	0	0	0	0	0	0	1	1	1	1
	1750	0	0	0	0	0	0	1	1	1	1
	2000	0	0	0	0	0	0	1	1	1	1

* Types RHH, RHW, and RHW-2 without outer covering.

TABLE C3 Maximum Number of Conductors and Fixture Wires in Flexible Metal Conduit (continued)

Type Letters	*Conductor Size AWG/kcmil*	*Conduit Trade Sizes in Inches*									
		½	*¾*	*1*	*1¼*	*1½*	*2*	*2½*	*3*	*3½*	*4*
THHN, THWN, THWN-2	14	13	22	33	52	76	134	202	291	396	518
	12	9	16	24	38	56	98	147	212	289	378
	10	6	10	15	24	35	62	93	134	182	238
	8	3	6	9	14	20	35	53	77	105	137
	6	2	4	6	10	14	25	38	55	76	99
	4	1	2	4	6	9	16	24	34	46	61
	3	1	1	3	5	7	13	20	29	39	51
	2	1	1	3	4	6	11	17	24	33	43
	1	1	1	1	3	4	8	12	18	24	32
	1/0	1	1	1	2	4	7	10	15	20	27
	2/0	0	1	1	1	3	6	9	12	17	22
	3/0	0	1	1	1	2	5	7	10	14	18
	4/0	0	1	1	1	1	4	6	8	12	15
	250	0	0	1	1	1	3	5	7	9	12
	300	0	0	1	1	1	3	4	6	8	11
	350	0	0	1	1	1	2	3	5	7	9

	400	0	0	0	1	1	1	3	5	6	8
	500	0	0	0	1	1	1	2	4	5	7
	600	0	0	0	0	1	1	1	3	4	5
	700	0	0	0	0	1	1	1	3	4	5
	750	0	0	0	0	1	1	1	2	3	4
	800	0	0	0	0	1	1	1	2	3	4
	900	0	0	0	0	0	1	1	1	3	4
	1000	0	0	0	0	0	1	1	1	3	3
FEP, FEBP,	14	12	21	32	51	74	130	196	282	385	502
PFA, PFAH,	12	9	15	24	37	54	95	143	206	281	367
TFE	10	6	11	17	26	39	68	103	148	201	263
	8	4	6	10	15	22	39	59	85	115	151
	6	2	4	7	11	16	28	42	60	82	107
	4	1	3	5	7	11	19	29	42	57	75
	3	1	2	4	6	9	16	24	35	48	62
	2	1	1	3	5	7	13	20	29	39	51
PFA, PFAH, TFE	1	1	1	2	3	5	9	14	20	27	36

TABLE C3 Maximum Number of Conductors and Fixture Wires in Flexible Metal Conduit (continued)

Type Letters	Conductor Size AWG/kcmil	Conduit Trade Sizes in Inches									
		½	¾	1	1¼	1½	2	2½	3	3½	4
PFA, PFAH, TFE, Z	1/0	1	1	1	3	4	8	11	17	23	30
	2/0	1	1	1	2	3	6	9	14	19	24
	3/0	0	1	1	1	3	5	8	11	15	20
	4/0	0	1	1	1	2	4	6	9	13	16
Z	14	15	25	39	61	89	157	236	340	463	605
	12	11	18	28	43	63	111	168	241	329	429
	10	6	11	17	26	39	68	103	148	201	263
	8	4	7	11	17	24	43	65	93	127	166
	6	3	5	7	12	17	30	45	65	89	117
	4	1	3	5	8	12	21	31	45	61	80
	3	1	2	4	6	8	15	23	33	45	58
	2	1	1	3	5	7	12	19	27	37	49
	1	1	1	2	4	6	10	15	22	30	39
XHH, XHHW, XHHW-2, ZW	14	9	15	23	36	53	94	141	203	277	361
	12	7	11	18	28	41	72	108	156	212	277
	10	5	8	13	21	30	54	81	116	158	207
	8	3	5	7	11	17	30	45	64	88	115
	6	1	3	5	8	12	22	33	48	65	85
	4	1	2	4	6	9	16	24	34	47	61
	3	1	1	3	5	7	13	20	29	40	52
	2	1	1	3	4	6	11	17	24	33	44

XHH, SHHW, XHHW-2	1	1	1	1	3	5	8	13	18	25	32
	1/0	1	1	1	2	4	7	10	15	21	27
	2/0	0	1	1	2	3	6	9	13	17	23
	3/0	0	1	1	1	3	5	7	10	14	19
	4/0	0	1	1	1	2	4	6	9	12	15
	250	0	0	1	1	1	3	5	7	10	13
	300	0	0	1	1	1	3	4	6	8	11
	350	0	0	1	1	1	2	4	5	7	9
	400	0	0	0	1	1	1	3	5	6	8
	500	0	0	0	1	1	1	3	4	5	7
	600	0	0	0	0	1	1	1	3	4	5
	700	0	0	0	0	1	1	1	3	4	5
	750	0	0	0	0	1	1	1	2	3	4
	800	0	0	0	0	1	1	1	2	3	4
	900	0	0	0	0	0	1	1	1	3	4
	1000	0	0	0	0	0	1	1	1	3	3
	1250	0	0	0	0	0	1	1	1	1	3
	1500	0	0	0	0	0	1	1	1	1	2
	1750	0	0	0	0	0	0	1	1	1	1
	2000	0	0	0	0	0	0	1	1	1	1

TABLE C3 Maximum Number of Conductors and Fixture Wires in Flexible Metal Conduit (continued)

Fixture Wires

Type Letters	*Conductor Size AWG/kcmil*	*Conduit Trade Size in Inches*					
		½	*¾*	*1*	*1¼*	*1½*	*2*
FFH-2, RFH-2, RFHH-3	18 16	8 7	14 12	22 19	35 29	51 43	90 76
SF-2, SFF-2	18 16 14	11 9 7	18 15 12	28 23 19	44 36 29	64 53 43	113 94 76
SF-1, SFF-1	18	19	32	50	78	114	201
AF, RFH-1, RFHH-2, TF, TFF, XF, XFF	18	14	24	37	58	84	148
AF, RFHH-2, TF, TFF, XF, XFF	16	11	19	30	47	68	120
AF, XF, XFF	14	9	15	23	36	53	94
TFN, TFFN	18 16	23 17	38 29	59 45	93 71	135 103	237 181
PF, PFF, PGF, PGFF, PAF, PTF, PTFF, PAFF	18 16 14	22 17 12	36 28 21	56 43 32	88 68 51	128 99 74	225 174 130

TABLE C3 (continued)

Type Letters	Conductor Size AWG/kcmil	Conduit Trade Size in Inches					
		½	¾	1	1¼	1½	2
ZF, ZFF,	18	28	47	72	113	165	290
ZHF, HF,	16	20	35	53	83	121	214
HFF	14	15	25	39	61	89	157
KF-2,	18	41	68	105	164	239	421
KFF-2	16	28	48	74	116	168	297
	14	19	33	51	80	116	204
	12	13	23	35	55	80	140
	10	9	15	23	36	53	94
KF-1,	18	48	82	125	196	285	503
KFF-1	16	34	57	88	138	200	353
	14	23	38	59	93	135	237
	12	15	25	39	61	89	157
	10	10	16	25	40	58	103
AF, XF,	12	5	8	12	19	28	50
XFF	10	4	6	10	15	22	39

Note: This table is for concentric stranded conductors only. For compact stranded conductors, Table C3A shall be used.

TABLE C3A Maximum Number of Compact Conductors in Flexible Metal Conduit

(Based on Table 1, Chapter 9)

Type Letters	*Conductor Size AWG/kcmil*	*Conduit Trade Sizes in Inches*									
		½	*¾*	*1*	*1¼*	*1½*	*2*	*2½*	*3*	*3½*	*4*
THW,	8	2	4	6	10	14	25	38	55	75	98
THHW,	6	1	3	5	7	11	20	29	43	58	76
THW-2	4	1	2	3	5	8	15	22	32	43	57
	2	1	1	2	4	6	11	16	23	32	42
	1	1	1	1	3	4	7	11	16	22	29
	1/0	1	1	1	2	3	6	10	14	19	25
	2/0	0	1	1	1	3	5	8	12	16	21
	3/0	0	1	1	1	2	4	7	10	14	18
	4/0	0	1	1	1	1	4	6	8	11	15
	250	0	0	1	1	1	3	4	7	9	12
	300	0	0	1	1	1	2	4	6	8	10
	350	0	0	1	1	1	2	3	5	7	9
	400	0	0	0	1	1	1	3	5	6	8
	500	0	0	0	1	1	1	3	4	5	7
	600	0	0	0	0	1	1	1	3	4	6
	700	0	0	0	0	1	1	1	3	4	5
	750	0	0	0	0	1	1	1	2	3	5
	1000	0	0	0	0	0	1	1	1	3	4

THHN,	8	—	—	—	—	—	—	—	—	—	—
THWN,	6	3	4	7	11	16	29	43	62	85	111
THWN-2	4	1	3	4	7	10	18	27	38	52	69
	2	1	1	3	5	7	13	19	28	38	49
	1	1	1	2	3	5	9	14	21	28	37
	1/0	1	1	1	3	4	8	12	17	24	31
	2/0	1	1	1	2	4	6	10	14	20	26
	3/0	0	1	1	1	3	5	8	12	17	22
	4/0	0	1	1	1	2	4	7	10	14	18
	250	0	1	1	1	1	3	5	8	11	14
	300	0	0	1	1	1	3	5	7	9	12
	350	0	0	1	1	1	3	4	6	8	10
	400	0	0	1	1	1	2	3	5	7	9
	500	0	0	0	1	1	1	3	4	6	8
	600	0	0	0	1	1	1	2	3	5	6
	700	0	0	0	0	1	1	1	3	4	6
	750	0	0	0	0	1	1	1	3	4	5
	1000	0	0	0	0	0	1	1	1	3	4

TABLE C3A Maximum Number of Compact Conductors in Flexible Metal Conduit

Type Letters	Conductor Size AWG/kcmil	Conduit Trade Sizes in Inches									
		½	¾	1	1¼	1½	2	2½	3	3½	4
XHHW, XHHW-2	8	3	5	8	13	19	33	50	71	97	127
	6	2	4	6	9	14	24	37	53	72	95
	4	1	3	4	7	10	18	27	38	52	69
	2	1	1	3	5	7	13	19	28	38	49
	1	1	1	2	3	5	9	14	21	28	37
	1/0	1	1	1	3	4	8	12	17	24	31
	2/0	1	1	1	2	4	7	10	15	20	26
	3/0	0	1	1	1	3	5	8	12	17	22
	4/0	0	1	1	1	2	4	7	10	14	18

	250	0	1	1	1	1	4	5	8	11	14
	300	0	0	1	1	1	3	5	7	9	12
	350	0	0	1	1	1	3	4	6	8	11
	400	0	0	1	1	1	2	4	5	7	10
	500	0	0	0	1	1	1	3	4	6	8
	600	0	0	0	1	1	1	2	3	5	6
	700	0	0	0	0	1	1	1	3	4	6
	750	0	0	0	0	1	1	1	3	4	5
	1000	0	0	0	0	1	1	1	2	3	4

Definition: Compact stranding is the result of a manufacturing process where the standard conductor is compressed to the extent that the interstices (voids between strand wires) are virtually eliminated.

TABLE C4 Maximum Number of Conductors and Fixture Wires in Intermediate Metal Conduit
(Based on Table 1, Chapter 9)

Type Letters	*Conductor Size AWG/kcmil*	*Conduit Trade Sizes in Inches*									
		½	*¾*	*1*	*1¼*	*1½*	*2*	*2½*	*3*	*3½*	*4*
RH	14	6	11	18	31	42	69	98	151	202	261
	12	5	9	14	25	34	56	79	122	163	209
RHH, RHW, RHW-2	14	4	8	13	22	30	49	70	108	144	136
	12	4	6	11	18	25	41	58	89	120	154
RHH, RHH, RHW, RHW-2	10	3	5	8	15	20	33	47	72	97	124
	8	1	3	4	8	10	17	24	38	50	65
	6	1	1	3	6	8	14	19	30	40	52
	4	1	1	3	5	6	11	15	23	31	41
	3	1	1	2	4	6	9	13	21	28	36
	2	1	1	1	3	5	8	11	18	24	31
	1	0	1	1	2	3	5	7	12	16	20
	1/0	0	1	1	1	3	4	6	10	14	18
	2/0	0	1	1	1	2	4	6	9	12	15
	3/0	0	0	1	1	1	3	5	7	10	13
	4/0	0	0	1	1	1	3	4	6	9	11

	250	0	0	1	1	1	1	3	5	6	8
	300	0	0	0	1	1	1	3	4	6	7
	350	0	0	0	1	1	1	2	4	5	7
	400	0	0	0	1	1	1	2	3	5	6
	500	0	0	0	1	1	1	1	3	4	5
	600	0	0	0	0	1	1	1	2	3	4
	700	0	0	0	0	1	1	1	2	3	4
	750	0	0	0	0	1	1	1	1	3	4
	800	0	0	0	0	0	1	1	1	3	3
	900	0	0	0	0	0	1	1	1	2	3
	1000	0	0	0	0	0	1	1	1	2	3
	1250	0	0	0	0	0	1	1	1	1	2
	1500	0	0	0	0	0	0	1	1	1	1
	1750	0	0	0	0	0	0	1	1	1	1
	2000	0	0	0	0	0	0	1	1	1	1
TW	14	10	17	27	47	64	104	147	228	304	392
	12	7	13	21	36	49	80	113	175	234	301
	10	5	9	15	27	36	59	84	130	174	224
	8	3	5	8	15	20	33	47	72	97	124

TABLE C4 Maximum Number of Conductors and Fixture Wires in Intermediate Metal Conduit (continued)

Type Letters	*Conductor Size AWG/kcmil*	*Conduit Trade Sizes in Inches*									
		½	*¾*	*1*	*1¼*	*1½*	*2*	*2½*	*3*	*3½*	*4*
RHH*, RHW*, RHW-2*, THHW, THW, THW-2	14	6	11	18	31	42	69	98	151	202	261
RHH*, RHW*, RHW-2*, THHW, THW	12	5	9	14	25	34	56	79	122	163	209
	10	4	7	11	19	26	43	61	95	127	163
RHH*, RHW*, RHW-2*, THHW, THW, THW-2	8	2	4	7	12	16	26	37	57	76	98
RHH*, RHW*, RHW-2*, TW, THW, THHW, THW-2	6	1	3	5	9	12	20	28	43	58	75
	4	1	2	4	6	9	15	21	32	43	56
	3	1	1	3	6	8	13	18	28	37	48
	2	1	1	3	5	6	11	15	23	31	41
	1	1	1	1	3	4	7	11	16	22	28

1/0	1	1	1	3	4	6	9	14	19	24
2/0	0	1	1	2	3	5	8	12	16	20
3/0	0	1	1	1	3	4	6	10	13	17
4/0	0	1	1	1	2	4	5	8	11	14
250	0	0	1	1	1	3	4	7	9	12
300	0	0	1	1	1	2	4	6	8	10
350	0	0	1	1	1	2	3	5	7	9
400	0	0	0	1	1	1	3	4	6	8
500	0	0	0	1	1	1	2	4	5	7
600	0	0	0	1	1	1	1	3	4	5
700	0	0	0	0	1	1	1	3	4	5
750	0	0	0	0	1	1	1	2	3	4
800	0	0	0	0	1	1	1	2	3	4
900	0	0	0	0	1	1	1	2	3	4
1000	0	0	0	0	0	1	1	1	3	3
1250	0	0	0	0	0	1	1	1	1	3
1500	0	0	0	0	0	1	1	1	1	2
1750	0	0	0	0	0	0	1	1	1	1
2000	0	0	0	0	0	0	1	1	1	1

* Types RHH, RHW, and RHW-2 without outer covering.

TABLE C4 Maximum Number of Conductors and Fixture Wires in Intermediate Metal Conduit (continued)

Type Letters	*Conductor Size AWG/kcmil*	*Conduit Trade Sizes in Inches*									
		½	*¾*	*1*	*1¼*	*1½*	*2*	*2½*	*3*	*3½*	*4*
THHN, THWN, THWN-2	14	14	24	39	68	91	149	211	326	436	562
	12	10	17	29	49	67	109	154	238	318	410
	10	6	11	18	31	42	68	97	150	200	258
	8	3	6	10	18	24	39	56	86	115	149
	6	2	4	7	13	17	28	40	62	83	107
	4	1	3	4	8	10	17	25	38	51	66
	3	1	2	4	6	9	15	21	32	43	56
	2	1	1	3	5	7	12	17	27	36	47
	1	1	1	2	4	5	9	13	20	27	35
	1/0	1	1	1	3	4	8	11	17	23	29
	2/0	1	1	1	3	4	6	9	14	19	24
	3/0	0	1	1	2	3	5	7	12	16	20
	4/0	0	1	1	1	2	4	6	9	13	17
	250	0	0	1	1	1	3	5	8	10	13
	300	0	0	1	1	1	3	4	7	9	12
	350	0	0	1	1	1	2	4	6	8	10

	400	0	0	1	1	1	2	3	5	7	9
	500	0	0	0	1	1	1	3	4	6	7
	600	0	0	0	1	1	1	2	3	5	6
	700	0	0	0	1	1	1	1	3	4	5
	750	0	0	0	1	1	1	1	3	4	5
	800	0	0	0	0	1	1	1	3	4	5
	900	0	0	0	0	1	1	1	2	3	4
	1000	0	0	0	0	1	1	1	2	3	4
FEP, FEPB, PFA, PFAH, TFE	14	13	23	38	66	89	145	205	317	423	545
	12	10	17	28	48	65	106	150	231	309	398
	10	7	12	20	34	46	76	107	166	221	285
	8	4	7	11	19	26	43	61	95	127	163
	6	3	5	8	14	19	31	44	67	90	116
	4	1	3	5	10	13	21	30	47	63	81
	3	1	3	4	8	11	18	25	39	52	68
	2	1	2	4	6	9	15	21	32	43	56
PFA, PFAH, TFE	1	1	1	2	4	6	10	14	22	30	39

TABLE C4 Maximum Number of Conductors and Fixture Wires in Intermediate Metal Conduit (continued)

Type Letters	Conductor Size AWG/kcmil	Conduit Trade Sizes in Inches									
		½	¾	1	1¼	1½	2	2½	3	3½	4
PFA, PFAH, TFE, Z	1/0	1	1	1	4	5	8	12	19	25	32
	2/0	1	1	1	3	4	7	10	15	21	27
	3/0	0	1	1	2	3	6	8	13	17	22
	4/0	0	1	1	1	3	5	7	10	14	18
Z	14	16	28	46	79	107	175	247	381	510	657
	12	11	20	32	56	76	124	175	271	362	466
	10	7	12	20	34	46	76	107	166	221	285
	8	4	7	12	21	29	48	68	105	140	180
	6	3	5	9	15	20	33	47	73	98	127
	4	1	3	6	10	14	23	33	50	67	87
	3	1	2	4	7	10	17	24	37	49	63
	2	1	1	3	6	8	14	20	30	41	53
	1	1	1	3	5	7	11	16	25	33	43
XHH, XHHW, XHHW-2, ZW	14	10	17	27	47	64	104	147	228	304	392
	12	7	13	21	36	49	80	113	175	234	301
	10	5	9	15	27	36	59	84	130	174	224
	8	3	5	8	15	20	33	47	72	97	124
	6	1	4	6	11	15	24	35	53	71	92

	4	1	3	4	8	11	18	25	39	52	67
	3	1	2	4	7	9	15	21	33	44	56
	2	1	1	3	5	7	12	18	27	37	47
XHH, XHHW, XHHW-2	1	1	1	2	4	5	9	13	20	27	35
	1/0	1	1	1	3	5	8	11	17	23	30
	2/0	1	1	1	3	4	6	9	14	19	25
	3/0	0	1	1	2	3	5	7	12	16	20
	4/0	0	1	1	1	2	4	6	10	13	17
	250	0	0	1	1	1	3	5	8	11	14
	300	0	0	1	1	1	3	4	7	9	12
	350	0	0	1	1	1	3	4	6	8	10
	400	0	0	1	1	1	2	3	5	7	9
	500	0	0	0	1	1	1	3	4	6	8
	600	0	0	0	1	1	1	2	3	5	6
	700	0	0	0	1	1	1	1	3	4	5
	750	0	0	0	1	1	1	1	3	4	5
	800	0	0	0	0	1	1	1	3	4	5
	900	0	0	0	0	1	1	1	2	3	4
	1000	0	0	0	0	1	1	1	2	3	4
	1250	0	0	0	0	0	1	1	1	2	3
	1500	0	0	0	0	0	1	1	1	1	2
	1750	0	0	0	0	0	1	1	1	1	2
	2000	0	0	0	0	0	0	1	1	1	1

TABLE C4 Maximum Number of Conductors and Fixture Wires in Intermediate Metal Conduit (continued)

Fixture Wires

Type Letters	*Conductor Size AWG/kcmil*	*Conduit Trade Sizes in Inches*					
		½	*¾*	*1*	*1¼*	*1½*	*2*
FHH-2, RFH-2, RFHH-3	18 16	9 8	16 13	26 22	45 38	61 51	100 84
SF-2, SFF-2	18 16 14	12 10 8	20 17 13	33 27 22	57 47 38	77 64 51	126 104 84
SF-1, SFF-1	18	21	36	59	101	137	223
AF, RFH-1, RFHH-2, TF, TFF, XF, XFF	18	15	26	43	75	101	165
AF, RFH-2, TF, TFF, XF, XFF	16	12	21	35	60	81	133
AF, XF, XFF	14	10	17	27	47	64	104
TFN, TFFN	18 16	25 19	42 32	69 53	119 91	161 123	264 201
PF, PFF, PGF, PGFF, PAF, PTF, PTFF, PAFF	18 16 14	23 18 13	40 31 23	66 51 38	113 87 66	153 118 89	250 193 145

TABLE C4 (continued)

Type Letters	Conductor Size AWG/kcmil	Conduit Trade Sizes in Inches					
		½	¾	1	1¼	1½	2
ZF, ZFF,	18	30	52	85	146	197	322
ZHF, HF,	16	22	38	63	108	145	238
HFF	14	16	28	46	79	107	175
KF-2,	18	44	75	123	212	287	468
KFF-2	16	31	53	87	149	202	330
	14	21	36	60	103	139	227
	12	14	25	41	70	95	156
	10	10	17	27	47	64	104
KF-1,	18	52	90	147	253	342	558
KFF-1	16	37	63	103	178	240	392
	14	25	42	69	119	161	264
	12	16	28	46	79	107	175
	10	10	18	30	52	70	114
AX, XF,	12	5	9	14	25	34	56
XFF	10	4	7	11	19	26	43

Note: This tabe is for concentric stranded conductors only. For compact stranded conductors, Table C4A shall be used.

TABLE C4A Maximum Number of Compact Conductors in Intermediate Metal Conduit
(Based on Table 1, Chapter 9)

Type Letters	*Conductor Size AWG/kcmil*	*Conduit Trade Sizes in Inches*									
		½	*¾*	*1*	*1¼*	*1½*	*2*	*2½*	*3*	*3½*	*4*
THW, THW-2, THHW	8	2	4	7	13	17	28	40	62	83	107
	6	1	3	6	10	13	22	31	48	64	82
	4	1	2	4	7	10	16	23	36	48	62
	2	1	1	3	5	7	12	17	26	35	45
	1	1	1	1	4	5	8	12	18	25	32
	1/0	1	1	1	3	4	7	10	16	21	27
	2/0	0	1	1	3	4	6	9	13	18	23
	3/0	0	1	1	2	3	5	7	11	15	20
	4/0	0	1	1	1	2	4	6	9	13	16
	250	0	0	1	1	1	3	5	7	10	13
	300	0	0	1	1	1	3	4	6	9	11
	350	0	0	1	1	1	2	4	6	8	10
	400	0	0	1	1	1	2	3	5	7	9
	500	0	0	0	1	1	1	3	4	6	8
	600	0	0	0	1	1	1	2	3	5	6
	700	0	0	0	1	1	1	1	3	4	5
	750	0	0	0	1	1	1	1	3	4	5
	1000	0	0	0	0	1	1	1	2	3	4

THHN,	8	—	—	—	—	—	—	—	—	—	—
THWN,	6	3	5	8	14	19	32	45	70	93	120
THWN-2	4	1	3	5	9	12	20	28	43	58	74
	2	1	1	3	6	8	14	20	31	41	53
	1	1	1	3	5	6	10	15	23	31	40
	1/0	1	1	2	4	5	9	13	20	26	34
	2/0	1	1	1	3	4	7	10	16	22	28
	3/0	0	1	1	3	4	6	9	14	18	24
	4/0	0	1	1	2	3	5	7	11	15	19
	250	0	1	1	1	2	4	6	9	12	15
	300	0	0	1	1	1	3	5	7	10	13
	350	0	0	1	1	1	3	4	7	9	11
	400	0	0	1	1	1	2	4	6	8	10
	500	0	0	1	1	1	2	3	5	7	9
	600	0	0	0	1	1	1	2	4	5	7
	700	0	0	0	1	1	1	2	3	5	6
	750	0	0	0	1	1	1	1	3	4	6
	1000	0	0	0	0	1	1	1	2	3	4

TABLE C4A Maximum Number of Compact Conductors in Intermediate Metal Conduit (continued)

Type Letters	*Conductor Size AWG/kcmil*	*Conduit Trade Sizes in Inches*									
		½	*¾*	*1*	*1¼*	*1½*	*2*	*2½*	*3*	*3½*	*4*
XHHW, XHHW-2	8	3	6	9	16	22	37	52	80	107	138
	6	2	4	7	12	16	27	38	59	80	103
	4	1	3	5	9	12	20	28	43	58	74
	2	1	1	3	6	8	14	20	31	41	53
	1	1	1	3	5	6	10	15	23	31	40
	1/0	1	1	2	4	5	9	13	20	26	34
	2/0	1	1	1	3	4	7	11	17	22	29
	3/0	0	1	1	3	4	6	9	14	18	24
	4/0	0	1	1	2	3	5	7	11	15	20
	250	0	1	1	1	2	4	6	9	12	16
	300	0	0	1	1	1	3	5	8	10	13
	350	0	0	1	1	1	3	4	7	9	12
	400	0	0	1	1	1	3	4	6	8	11
	500	0	0	1	1	1	2	3	5	7	9
	600	0	0	0	1	1	1	2	4	5	7
	700	0	0	0	1	1	1	2	3	5	6
	750	0	0	0	1	1	1	1	3	4	6
	1000	0	0	0	0	1	1	1	2	3	4

Definition: Compact stranding is the result of a manufacturing process were the standard conductor is compressed to the extent that interstices (voids between strand wires) are virtually eliminated.

TABLE C5 Maximum Number of Conductors and Fixture Wires in Liquidtight Flexible Nonmetallic Conduit (Type FNMC-B**)

(Based on Table 1, Chapter 9)

Type Letters	Conductor Size AWG/kcmil	Conduit Trade Sizes in Inches						
		⅜	½	¾	1	1¼	1½	2
RH	14	3	6	10	16	29	38	62
	12	3	5	8	12	23	30	50
RHH, RHW, RHW-2	14	2	4	7	12	21	27	44
	12	1	3	6	10	17	22	36
RH, RHH, RHW, RHW-2	10	1	3	5	8	14	18	29
	8	1	1	2	4	7	9	15
	6	1	1	1	3	6	7	12
	4	0	1	1	2	4	6	9
	3	0	1	1	1	4	5	8
	2	0	1	1	1	3	4	7
	1	0	0	1	1	1	3	5
	1/0	0	0	1	1	1	2	4
	2/0	0	0	1	1	1	1	3
	3/0	0	0	0	1	1	1	3
	4/0	0	0	0	1	1	1	2
	250	0	0	0	0	1	1	1
	300	0	0	0	0	1	1	1
	350	0	0	0	0	1	1	1
	400	0	0	0	0	1	1	1
	500	0	0	0	0	1	1	1
	600	0	0	0	0	0	1	1
	700	0	0	0	0	0	0	1
	750	0	0	0	0	0	0	1
	800	0	0	0	0	0	0	1
	900	0	0	0	0	0	0	1
	1000	0	0	0	0	0	0	1
	1250	0	0	0	0	0	0	0
	1500	0	0	0	0	0	0	0
	1750	0	0	0	0	0	0	0
	2000	0	0	0	0	0	0	0

** Corresponds to Section 351-22(2).

TABLE C5 Maximum Number of Conductors and Fixture Wires in Liquidtight Flexible Nonmetallic Conduit (Type FNMC-B) (continued)**

Type Letters	*Conductor Size AWG/kcmil*	*Conduit Trade Sizes in Inches*						
		⅜	½	¾	1	1¼	1½	2
TW	14	5	9	15	25	44	57	93
	12	4	7	12	19	33	43	71
	10	3	5	9	14	25	32	53
	8	1	3	5	8	14	18	29
RHH*, RHW*, RHW-2*, THHW, THW, THW-2	14	3	6	10	16	29	38	62
RHH*, RHW*, RHW-2*, THHW, THW	12	3	5	8	13	23	30	50
	10	1	3	6	10	18	23	39
RHH*, RHW*, RHW-2*, THHW, THW, THW-2	8	1	1	4	6	11	14	23
RHH*, RHW*, RHW-2*, TW, THW, THHW, THW-2	6	1	1	3	5	8	11	18
	4	1	1	1	3	6	8	13
	3	1	1	1	3	5	7	11
	2	0	1	1	2	4	6	9
	1	0	1	1	1	3	4	7
	1/0	0	0	1	1	2	3	6
	2/0	0	0	1	1	2	3	5
	3/0	0	0	1	1	1	2	4
	4/0	0	0	0	1	1	1	3
	250	0	0	0	1	1	1	3
	300	0	0	0	1	1	1	2
	350	0	0	0	0	1	1	1
	400	0	0	0	0	1	1	1
	500	0	0	0	0	1	1	1
	600	0	0	0	0	1	1	1
	700	0	0	0	0	0	1	1
	750	0	0	0	0	0	1	1
	800	0	0	0	0	0	1	1
	900	0	0	0	0	0	0	1
	1000	0	0	0	0	0	0	1
	1250	0	0	0	0	0	0	1
	1500	0	0	0	0	0	0	0
	1750	0	0	0	0	0	0	0
	2000	0	0	0	0	0	0	0

* Types RHH, RHW, and RHW-2 without outer covering.

TABLE C5 Maximum Number of Conductors and Fixture Wires in Liquidtight Flexible Nonmetallic Conduit (Type FNMC-B**) (continued)

Type Letters	Conductor Size AWG/kcmil	Conduit Trade Sizes in Inches						
		⅜	½	¾	1	1¼	1½	2
THHN, THWN, THWN-2	14	8	13	22	36	63	81	133
	12	5	9	16	26	46	59	97
	10	3	6	10	16	29	37	61
	8	1	3	6	9	16	21	35
	6	1	2	4	7	12	15	25
	4	1	1	2	4	7	9	15
	3	1	1	1	3	6	8	13
	2	1	1	1	3	5	7	11
	1	0	1	1	1	4	5	8
	1/0	0	1	1	1	3	4	7
	2/0	0	0	1	1	2	3	6
	3/0	0	0	1	1	1	3	5
	4/0	0	0	1	1	1	2	4
	250	0	0	0	1	1	1	3
	300	0	0	0	1	1	1	3
	350	0	0	0	1	1	1	2
	400	0	0	0	0	1	1	1
	500	0	0	0	0	1	1	1
	600	0	0	0	0	1	1	1
	700	0	0	0	0	1	1	1
	750	0	0	0	0	0	1	1
	800	0	0	0	0	0	1	1
	900	0	0	0	0	0	1	1
	1000	0	0	0	0	0	0	1
FEP, FEPB, PFA, PFAH, TFE	14	7	12	21	35	61	79	129
	12	5	9	15	25	44	57	94
	10	4	6	11	18	32	41	68
	8	1	3	6	10	18	23	39
	6	1	2	4	7	13	17	27
	4	1	1	3	5	9	12	19
	3	1	1	2	4	7	10	16
	2	1	1	1	3	6	8	13
PFA, PFAH, TFE	1	0	1	1	2	4	5	9

** Corresponds to Section 351-22(2).

TABLE C5 Maximum Number of Conductors and Fixture Wires in Liquidtight Flexible Nonmetallic Conduit (Type FNMC-B**) (continued)

Type Letters	Conductor Size AWG/kcmil	Conduit Trade Sizes in Inches						
		3/8	1/2	3/4	1	1 1/4	1 1/2	2
PFA, PFAH, TFE, Z	1/0	0	1	1	1	3	4	7
	2/0	0	1	1	1	3	4	6
	3/0	0	0	1	1	2	3	5
	4/0	0	0	1	1	1	2	4
Z	14	9	15	26	42	73	95	156
	12	6	10	18	30	52	67	111
	10	4	6	11	18	32	41	68
	8	2	4	7	11	20	26	43
	6	1	3	5	8	14	18	30
	4	1	1	3	5	9	12	20
	3	1	1	2	4	7	9	15
	2	0	1	1	3	6	7	12
	1	0	1	1	2	5	6	10
XHH, XHHW, XHHW-2, ZW	14	5	9	15	25	44	57	93
	12	4	7	12	19	33	43	71
	10	3	5	9	14	25	32	53
	8	1	3	5	8	14	18	29
	6	1	1	3	6	10	13	22
	4	1	1	2	4	7	9	16
	3	1	1	1	3	6	8	13
	2	1	1	1	3	5	7	11
XHH, XHHW, XHHW-2	1	0	1	1	1	4	5	8
	1/0	0	1	1	1	3	4	7
	2/0	0	0	1	1	2	3	6
	3/0	0	0	1	1	1	3	5
	4/0	0	0	1	1	1	2	4
	250	0	0	0	1	1	1	3
	300	0	0	0	1	1	1	3
	350	0	0	0	1	1	1	2
	400	0	0	0	0	1	1	1
	500	0	0	0	0	1	1	1
	600	0	0	0	0	1	1	1
	700	0	0	0	0	1	1	1
	750	0	0	0	0	0	1	1
	800	0	0	0	0	0	1	1
	900	0	0	0	0	0	1	1
	1000	0	0	0	0	0	0	1
	1250	0	0	0	0	0	0	1
	1500	0	0	0	0	0	0	1
	1750	0	0	0	0	0	0	0
	2000	0	0	0	0	0	0	0

** Corresponds to Section 351-22(2).

TABLE C5 Maximum Number of Conductors and Fixture Wires in Liquidtight Flexible Nonmetallic Conduit (Type FNMC-B**) (continued)

Fixture Wires

Type Letters	*Conductor Size AWG/kcmil*	*Conduit Trade Sizes in Inches*						
		⅜	*½*	*¾*	*1*	*1¼*	*1½*	*2*
FFH-2, RFH-2	18	5	8	15	24	42	54	89
	16	4	7	12	20	35	46	75
SF-2, SFF-2	18	6	11	19	30	53	69	113
	16	5	9	15	25	44	57	93
	14	4	7	12	20	35	46	75
SF-1, SFF-1	18	11	19	33	53	94	122	199
AF, RFH-1, RFHH-2, TF, TFE, XF, XFF	18	8	14	24	39	69	90	147
AF, RFHH-2, TF, TFF, XF, XFF	16	7	11	20	32	56	72	119
AF, XF, XFF	14	5	9	15	25	44	57	93
TFN, TFFN	18	14	23	39	63	111	144	236
	16	10	17	30	48	85	110	180
PF, PFF, PGF, PGFF, PAF, PTF, PTFF, PAFF	18	13	21	37	60	105	136	223
	16	10	16	29	46	81	105	173
	14	7	12	21	35	61	79	129
HF, HFF, ZF, ZFF, ZHF	18	12	28	48	77	136	176	288
	16	12	20	35	57	100	129	212
	14	9	15	26	42	73	95	156
KF-2, KFF-2	18	24	40	70	112	197	255	418
	16	17	28	49	79	139	180	295
	14	12	19	34	54	95	123	202
	12	8	13	23	37	65	85	139
	10	5	9	15	25	44	57	93
KF-1, KFF-1	18	29	48	83	134	235	304	499
	16	20	34	58	94	165	214	350
	14	14	23	39	63	111	144	236
	12	9	15	26	42	73	95	156
	10	6	10	17	27	48	62	102
AF, XF, XFF	12	3	5	8	13	23	30	50
	10	1	3	6	10	18	23	39

** Corresponds to Section 351-22(2).

Note: This table is for concentric stranded conductors only. For compact stranded conductors, Table C5A should be used.

TABLE C5A Maximum Number of Compact Conductors in Liquidtight Flexible Nonmetallic Conduit (Type FNMC-B**)

(Based on Table 1, Chapter 9)

Type Letters	Conductor Size AWG/kcmil	Conduit Trade Sizes in Inches						
		3/8	1/2	3/4	1	1¼	1½	2
THW, THW-2, THHW	8	1	2	4	7	12	15	25
	6	1	1	3	5	9	12	19
	4	1	1	2	4	7	9	14
	2	1	1	1	3	5	6	11
	1	0	1	1	1	3	4	7
	1/0	0	1	1	1	3	4	6
	2/0	0	0	1	1	2	3	5
	3/0	0	0	1	1	1	3	4
	4/0	0	0	1	1	1	2	4
	250	0	0	0	1	1	1	3
	300	0	0	0	1	1	1	2
	350	0	0	0	1	1	1	2
	400	0	0	0	0	1	1	1
	500	0	0	0	0	1	1	1
	600	0	0	0	0	1	1	1
	700	0	0	0	0	1	1	1
	750	0	0	0	0	0	1	1
	1000	0	0	0	0	0	1	1
THHN, THWN, THWN-2	8	-	-	-	-	-	-	-
	6	1	2	4	7	13	17	28
	4	1	1	3	4	8	11	17
	2	1	1	1	3	6	7	12
	1	0	1	1	2	4	6	9
	1/0	0	1	1	1	4	5	8
	2/0	0	1	1	1	3	4	6
	3/0	0	0	1	1	2	3	5
	4/0	0	0	1	1	1	3	4
	250	0	0	1	1	1	1	3
	300	0	0	0	1	1	1	3
	350	0	0	0	1	1	1	2
	400	0	0	0	1	1	1	2
	500	0	0	0	0	1	1	1
	600	0	0	0	0	1	1	1
	700	0	0	0	0	1	1	1
	750	0	0	0	0	1	1	1
	1000	0	0	0	0	0	1	1

** Corresponds to Section 351-22(2).

TABLE C5A Maximum Number of Compact Conductors in Liquidtight Flexible Nonmetallic Conduit (Type FNMC-B) (continued)**

Type Letters	*Conductor Size AWG/kcmil*	*Conduit Trade Sizes in Inches*						
		3/8	*1/2*	*3/4*	*1*	*1¼*	*1½*	*2*
XHHW, XHHW-2	8	1	3	5	9	15	20	33
	6	1	2	4	6	11	15	24
	4	1	1	3	4	8	11	17
	2	1	1	1	3	6	7	12
	1	0	1	1	2	4	6	9
	1/0	0	1	1	1	4	5	8
	2/0	0	1	1	1	3	4	7
	3/0	0	0	1	1	2	3	5
	4/0	0	0	1	1	1	3	4
	250	0	0	1	1	1	1	3
	300	0	0	0	1	1	1	3
	350	0	0	0	1	1	1	3
	400	0	0	0	1	1	1	2
	500	0	0	0	0	1	1	1
	600	0	0	0	0	1	1	1
	700	0	0	0	0	1	1	1
	750	0	0	0	0	1	1	1
	1000	0	0	0	0	0	1	1

Definition: Compact stranding is the result of a manufacturing process where the standard conductor is compressed to the extent that the interstices (voids between strand wires) are virtually eliminated.

TABLE C6 Maximum Number of Conductors and Fixture Wires in Liquidtight Flexible Nonmetallic Conduit (Type FNMC-A**)

(Based on Table 1, Chapter 9)

Type Letters	*Conductor Size AWG/kcmil*	*Conduit Trade Sizes in Inches*						
		⅜	*½*	*¾*	*1*	*1¼*	*1½*	*2*
RH	14	3	6	10	16	28	38	64
	12	3	4	8	13	23	31	51
RHH, RHW, RHW-2	14	2	4	7	11	20	27	45
	12	1	3	6	9	17	23	38
RH, RHH, RHW, RHW-2	10	1	3	5	8	13	18	30
	8	1	1	2	4	7	9	16
	6	1	1	1	3	5	7	13
	4	0	1	1	2	4	6	10
	3	0	1	1	1	4	5	8
	2	0	1	1	1	3	4	7
	1	0	0	1	1	1	3	5
	1/0	0	0	1	1	1	2	4
	2/0	0	0	1	1	1	1	4
	3/0	0	0	0	1	1	1	3
	4/0	0	0	0	1	1	1	3
	250	0	0	0	0	1	1	1
	300	0	0	0	0	1	1	1
	350	0	0	0	0	1	1	1
	400	0	0	0	0	1	1	1
	500	0	0	0	0	0	1	1
	600	0	0	0	0	0	1	1
	700	0	0	0	0	0	0	1
	750	0	0	0	0	0	0	1
	800	0	0	0	0	0	0	1
	900	0	0	0	0	0	0	1
	1000	0	0	0	0	0	0	1
	1250	0	0	0	0	0	0	0
	1500	0	0	0	0	0	0	0
	1750	0	0	0	0	0	0	0
	2000	0	0	0	0	0	0	0

** Corresponds to Section 351-22(1).

TABLE C6 Maximum Number of Conductors and Fixture Wires in Liquidtight Flexible Nonmetallic Conduit (Type FNMC-A**) (continued)

Type Letters	Conductor Size AWG/kcmil	Conduit Trade Sizes in Inches						
		⅜	½	¾	1	1¼	1½	2
TW	14	5	9	15	24	43	58	96
	12	4	7	12	19	33	44	74
	10	3	5	9	14	24	33	55
	8	1	3	5	8	13	18	30
RHH*, RHW*, RHW-2*, THHW, THW, THW-2	14	3	6	10	16	28	38	64
RHH*, RHW*, RHW-2*, THHW, THW	12	3	4	8	13	23	31	51
	10	1	3	6	10	18	24	40
RHH*, RHW*, RHW-2, THHW, THW, THW-2	8	1	1	4	6	10	14	24
RHH*, RHW*, RHW-2*, TW, THW, THHW, THW-2	6	1	1	3	4	8	11	18
	4	1	1	1	3	6	8	13
	3	1	1	1	3	5	7	11
	2	0	1	1	2	4	6	10
	1	0	1	1	1	3	4	7
	1/0	0	0	1	1	2	3	6
	2/0	0	0	1	1	1	3	5
	3/0	0	0	1	1	1	2	4
	4/0	0	0	0	1	1	1	3
	250	0	0	0	1	1	1	3
	300	0	0	0	1	1	1	2
	350	0	0	0	0	1	1	1
	400	0	0	0	0	1	1	1
	500	0	0	0	0	1	1	1
	600	0	0	0	0	1	1	1
	700	0	0	0	0	0	1	1
	750	0	0	0	0	0	1	1
	800	0	0	0	0	0	1	1
	900	0	0	0	0	0	0	1
	1000	0	0	0	0	0	0	1
	1250	0	0	0	0	0	0	1
	1500	0	0	0	0	0	0	1
	1750	0	0	0	0	0	0	0
	2000	0	0	0	0	0	0	0

* Types RHH, RHW, and RHW-2 without outer covering.

** Corresponds to Section 351-22(1).

TABLE C6 Maximum Number of Conductors and Fixture Wires in Liquidtight Flexible Nonmetallic Conduit (Type FNMC-A) (continued)**

Type Letters	*Conductor Size AWG/kcmil*	*Conduit Trade Sizes in Inches*						
		3/8	*1/2*	*3/4*	*1*	*1 1/4*	*1 1/2*	*2*
THHN, THWN, THWN-2	14	8	13	22	35	62	83	137
	12	5	9	16	25	45	60	100
	10	3	6	10	16	28	38	63
	8	1	3	6	9	16	22	36
	6	1	2	4	6	12	16	26
	4	1	1	2	4	7	9	16
	3	1	1	1	3	6	8	13
	2	1	1	1	1	4	5	8
	1/0	0	1	1	1	3	4	7
	2/0	0	0	1	1	2	3	6
	3/0	0	0	1	1	1	3	5
	4/0	0	0	1	1	1	2	4
	250	0	0	0	1	1	1	3
	300	0	0	0	1	1	1	3
	350	0	0	0	1	1	1	2
	400	0	0	0	0	1	1	1
	500	0	0	0	0	1	1	1
	600	0	0	0	0	1	1	1
	750	0	0	0	0	0	1	1
	800	0	0	0	0	0	1	1
	900	0	0	0	0	0	1	1
	1000	0	0	0	0	0	0	1
FEP, FEPB, PFA, PFAH, TFE	14	7	12	21	34	60	80	133
	12	5	9	15	25	44	59	97
	10	4	6	11	18	31	42	70
	8	1	3	6	10	18	24	40
	6	1	2	4	7	13	17	28
	4	1	1	3	5	9	12	20
	3	1	1	2	4	7	10	16
	2	1	1	1	3	6	8	13
PFA, PFAH, TFE	1	0	1	1	2	4	5	9

** Corresponds to Section 351-22(1).

TABLE C6 Maximum Number of Conductors and Fixture Wires in Liquidtight Flexible Nonmetallic Conduit (Type FNMC-A**) (continued)

Type Letters	Conductor Size AWG/kcmil	Conduit Trade Sizes in Inches						
		3/8	1/2	3/4	1	1¼	1½	2
PFA, PFAH, TFE, Z	1/0	0	1	1	1	3	5	8
	2/0	0	1	1	1	3	4	6
	3/0	0	0	1	1	2	3	5
	4/0	0	0	1	1	1	2	4
Z	14	9	15	25	41	72	97	161
	12	6	10	18	29	51	69	114
	10	4	6	11	18	31	42	70
	8	2	4	7	11	20	26	44
	6	1	3	5	8	14	18	31
	4	1	1	3	5	9	13	21
	3	1	1	2	4	7	9	15
	2	1	1	1	3	6	8	13
	1	1	1	1	2	4	6	10
XHH, XHHW, XHHW-2, ZW	14	5	9	15	24	43	58	96
	12	4	7	12	19	33	44	74
	10	3	5	9	14	24	33	55
	8	1	3	5	8	13	18	30
	6	1	1	3	5	10	13	22
	4	1	1	2	4	7	10	16
	3	1	1	1	3	6	8	14
	2	1	1	1	3	5	7	11
XHH, XHHW, XHHW-2	1	0	1	1	1	4	5	8
	1/0	0	1	1	1	3	4	7
	2/0	0	0	1	1	2	3	6
	3/0	0	0	1	1	1	3	5
	4/0	0	0	1	1	1	2	4
	250	0	0	0	1	1	1	3
	300	0	0	0	1	1	1	3
	350	0	0	0	1	1	1	2
	400	0	0	0	0	1	1	1
	500	0	0	0	0	1	1	1
	600	0	0	0	0	1	1	1
	700	0	0	0	0	1	1	1
	750	0	0	0	0	0	1	1
	800	0	0	0	0	0	1	1
	900	0	0	0	0	0	1	1
	1000	0	0	0	0	0	0	1
	1250	0	0	0	0	0	0	1
	1500	0	0	0	0	0	0	1
	1750	0	0	0	0	0	0	0
	2000	0	0	0	0	0	0	0

** Corresponds to Section 351-22(1).

TABLE C6 Maximum Number of Conductors and Fixture Wires in Liquidtight Flexible Nonmetallic Conduit (Type FNMC-A**) (continued)

Fixture Wires

Type Letters	Conductor Size AWG/kcmil	Conduit Trade Sizes in Inches						
		⅜	½	¾	1	1¼	1½	2
FFH-2, RFH-2, RFHH-3	18 16	5 4	8 7	14 12	23 20	41 35	55 47	92 77
SF-2, SFF-2	18 16 14	6 5 4	11 9 7	18 15 12	29 24 20	52 43 35	70 58 47	116 96 77
SF-1, SFF-1	18	12	19	33	52	92	124	205
AF, RFH-1, RFHH-2, TF, TFF, XF, XFF	18	8	14	24	39	68	91	152
AF, RFHH-2, TF, TFF, XF, XFF	16	7	11	19	31	55	74	122
AF, XF, XFF	14	5	9	15	24	43	58	96
TFN, TFFN	18 16	14 10	22 17	39 29	62 47	109 83	146 112	243 185
PF, PFF, PGF, PGFF, PAF, PTF, PTFF, PAFF	18 16 14	13 10 7	21 16 12	37 28 21	59 45 34	103 80 60	139 107 80	230 178 133

TABLE C6 (continued)

Type Letters	Conductor Size AWG/kcmil	Conduit Trade Sizes in Inches						
		⅜	½	¾	1	1¼	1½	2
HF, HFF,	18	17	27	47	76	133	179	297
ZF, ZFF,	16	12	20	35	56	98	132	219
ZHF	14	9	15	25	41	72	97	161
KF-2,	18	25	40	69	110	193	260	431
KFF-2	16	17	28	48	77	136	183	303
	14	12	19	33	53	94	126	209
	12	8	13	23	36	64	86	143
	10	5	9	15	24	43	58	96
KF-1,	18	29	48	82	131	231	310	514
KFF-1	16	21	33	57	92	162	218	361
	14	14	22	39	62	109	146	243
	12	9	15	25	41	72	97	161
	10	6	10	17	27	47	63	105
AF, XF,	12	3	4	8	13	23	31	51
XFF	10	1	3	6	10	18	24	40

** Corresponds to Section 351-22(1).

Note: This table is for concentric stranded conductors only. For compact stranded conductors, Table C7A should be used.

TABLE C6A Maximum Number of Compact Conductors in Liquidtight Flexible Nonmetallic Conduit (Type FNMC-A**)

(Based on Table 1, Chapter 9)

Type Letters	*Conductor Size AWG/kcmil*	*Conduit Trade Sizes in Inches*						
		3/8	*1/2*	*3/4*	*1*	*1 1/4*	*1 1/2*	*2*
THW, THW-2, THHW	8	1	2	4	6	11	16	26
	6	1	1	3	5	9	12	20
	4	1	1	2	4	7	9	15
	2	1	1	1	3	5	6	11
	1	0	1	1	1	3	4	8
	1/0	0	1	1	1	3	4	7
	2/0	0	0	1	1	2	3	5
	3/0	0	0	1	1	1	3	5
	4/0	0	0	1	1	1	2	4
	250	0	0	0	1	1	1	3
	300	0	0	0	1	1	1	3
	350	0	0	0	1	1	1	2
	400	0	0	0	0	1	1	1
	500	0	0	0	0	1	1	1
	600	0	0	0	0	1	1	1
	700	0	0	0	0	1	1	1
	750	0	0	0	0	0	1	1
	1000	0	0	0	0	0	1	1
THHN, THWN, THWN-2	8	-	-	-	-	-	-	-
	6	1	2	4	7	13	18	29
	4	1	1	3	4	8	11	18
	2	1	1	1	3	6	8	13
	1	0	1	1	2	4	6	10
	1/0	0	1	1	1	3	5	8
	2/0	0	1	1	1	3	4	7
	3/0	0	0	1	1	2	3	6
	4/0	0	0	1	1	1	3	5
	250	0	0	1	1	1	1	3
	300	0	0	0	1	1	1	3
	350	0	0	0	1	1	1	3
	400	0	0	0	1	1	1	2
	500	0	0	0	0	1	1	1
	600	0	0	0	0	1	1	1
	700	0	0	0	0	1	1	1
	750	0	0	0	0	1	1	1
	1000	0	0	0	0	0	1	1

** Corresponds to Section 351-22(1).

TABLE C6A Maximum Number of Compact Conductors in Liquidtight Flexible Nonmetallic Conduit (Type FNMC-A) (continued)

Type Letters	Conductor Size AWG/kcmil	Conduit Trade Sizes in Inches						
		⅜	½	¾	1	1¼	1½	2
XHHW, XHHW-2	8	1	3	5	8	15	20	34
	6	1	2	4	6	11	15	25
	4	1	1	3	4	8	11	18
	2	1	1	1	3	6	8	13
	1	0	1	1	2	4	6	10
	1/0	0	1	1	1	3	5	8
	2/0	0	1	1	1	3	4	7
	3/0	0	0	1	1	2	3	6
	4/0	0	0	1	1	1	3	5
	250	0	0	1	1	1	2	4
	300	0	0	0	1	1	1	3
	350	0	0	0	1	1	1	3
	400	0	0	0	1	1	1	2
	500	0	0	0	0	1	1	1
	600	0	0	0	0	1	1	1
	700	0	0	0	0	1	1	1
	750	0	0	0	0	1	1	1
	1000	0	0	0	0	0	1	1

Definition: Compact stranding is the result of a manufacturing process where the standard conductor is compressed to the extent that the interstices (voids between strand wires) are virtually eliminated.

TABLE C7 Maximum Number of Conductors and Fixture Wires in Liquidtight Flexible Metal Conduit
(Based on Table 1, Chapter 9)

Type Letters	*Conductor Size AWG/kcmil*	*Conduit Trade Sizes in Inches*									
		½	*¾*	*1*	*1¼*	*1½*	*2*	*2½*	*3*	*3½*	*4*
RH	14	6	10	16	29	38	62	93	143	186	243
	12	5	8	13	23	30	50	75	115	149	195
RHH, RHW, RHW-2	14	4	7	12	21	27	44	66	102	133	173
	12	3	6	10	17	22	36	55	84	110	144
RH, RHH, RHW, RHW-2	10	3	5	8	14	18	29	44	68	89	116
	8	1	2	4	7	9	15	23	36	46	61
	6	1	1	3	6	7	12	18	28	37	48
	4	1	1	2	4	6	9	14	22	29	38
	3	1	1	1	4	5	8	13	19	25	33
	2	1	1	1	3	4	7	11	17	22	29
	1	0	1	1	1	3	5	7	11	14	19

	1/0	0	1	1	1	2	4	6	10	13	16
	2/0	0	1	1	1	1	3	5	8	11	14
	3/0	0	0	1	1	1	3	4	7	9	12
	4/0	0	0	1	1	1	2	4	6	8	10
	250	0	0	0	1	1	1	3	4	6	8
	300	0	0	0	1	1	1	2	4	5	7
	350	0	0	0	1	1	1	2	3	5	6
	400	0	0	0	1	1	1	1	3	4	6
	500	0	0	0	1	1	1	1	3	4	5
	600	0	0	0	0	1	1	1	2	3	4
	700	0	0	0	0	0	1	1	1	3	3
	750	0	0	0	0	0	1	1	1	2	3
	800	0	0	0	0	0	1	1	1	2	3
	900	0	0	0	0	0	1	1	1	2	3
	1000	0	0	0	0	0	1	1	1	1	3
	1250	0	0	0	0	0	0	1	1	1	1
	1500	0	0	0	0	0	0	1	1	1	1
	1750	0	0	0	0	0	0	1	1	1	1
	2000	0	0	0	0	0	0	0	1	1	1

TABLE C7 Maximum Number of Conductors and Fixture Wires in Liquidtight Flexible Metal Conduit (continued)

Type Letters	*Conductor Size AWG/kcmil*	*Conduit Trade Sizes in Inches*									
		½	*¾*	*1*	*1¼*	*1½*	*2*	*2½*	*3*	*3½*	*4*
TW	14	9	15	25	44	57	93	140	215	280	365
	12	7	12	19	33	43	71	108	165	215	280
	10	5	9	14	25	32	53	80	123	160	209
	8	3	5	8	14	18	29	44	68	89	116
RHH*, RHW*, RHW-2*, THHW, THW, THW-2	14	6	10	16	29	38	62	93	143	186	243
RHH*, RHW*, RHW-2*, THHW, THW	12	5	8	13	23	30	50	75	115	149	195
	10	3	6	10	18	23	39	58	89	117	152
RHH*, RHW*, RHW-2*, THHW, THW, THW-2	8	1	4	6	11	14	23	35	53	70	91

RHH*, RHW*,	6	1	3	5	8	11	18	27	41	53	70
RHW-2*,	4	1	1	3	6	8	13	20	30	40	52
TW, THW,	3	1	1	3	5	7	11	17	26	34	44
THHW,	2	1	1	2	4	6	9	14	22	29	38
THW-2	1	1	1	1	3	4	7	10	15	20	26
	1/0	0	1	1	2	3	6	8	13	17	23
	2/0	0	1	1	2	3	5	7	11	15	19
	3/0	0	1	1	1	2	4	6	9	12	16
	4/0	0	0	1	1	1	3	5	8	10	13
	250	0	0	1	1	1	3	4	6	8	11
	300	0	0	1	1	1	2	3	5	7	9
	350	0	0	0	1	1	1	3	5	6	8
	400	0	0	0	1	1	1	3	4	6	7
	500	0	0	0	1	1	1	2	3	5	6
	600	0	0	0	1	1	1	1	3	4	5
	700	0	0	0	0	1	1	1	2	3	4
	750	0	0	0	0	1	1	1	2	3	4
	800	0	0	0	0	1	1	1	2	3	4
	900	0	0	0	0	0	1	1	1	3	3
	1000	0	0	0	0	0	1	1	1	2	3
	1250	0	0	0	0	0	1	1	1	1	2
	1500	0	0	0	0	0	0	1	1	1	2
	1750	0	0	0	0	0	0	1	1	1	1
	2000	0	0	0	0	0	0	1	1	1	1

* Types RHH, RHW, and RHW-2 without outer covering.

TABLE C7 Maximum Number of Conductors and Fixture Wires in Liquidtight Flexible Metal Conduit (continued)

Type Letters	*Conductor Size AWG/kcmil*	*Conduit Trade Sizes in Inches*									
		½	*¾*	*1*	*1¼*	*1½*	*2*	*2½*	*3*	*3½*	*4*
THHN, THWN, THWN-2	14	13	22	36	63	81	133	201	308	401	523
	12	9	16	26	46	59	97	146	225	292	381
	10	6	10	16	29	37	61	92	141	184	240
	8	3	6	9	16	21	35	53	81	106	138
	6	2	4	7	12	15	25	38	59	76	100
	4	1	2	4	7	9	15	23	36	47	61
	3	1	1	3	6	8	13	20	30	40	52
	2	1	1	3	5	7	11	17	26	33	44
	1	1	1	1	4	5	8	12	19	25	32
	1/0	1	1	1	3	4	7	10	16	21	27
	2/0	0	1	1	2	3	6	8	13	17	23
	3/0	0	1	1	1	3	5	7	11	14	19
	4/0	0	1	1	1	2	4	6	9	12	15
	250	0	0	1	1	1	3	5	7	10	12
	300	0	0	1	1	1	3	4	6	8	11
	350	0	0	1	1	1	2	3	5	7	9

	400	0	0	0	1	1	1	3	5	6	8
	500	0	0	0	1	1	1	2	4	5	7
	600	0	0	0	1	1	1	1	3	4	6
	700	0	0	0	1	1	1	1	3	4	5
	750	0	0	0	0	1	1	1	3	3	5
	800	0	0	0	0	1	1	1	2	3	4
	900	0	0	0	0	1	1	1	2	3	4
	1000	0	0	0	0	0	1	1	1	3	3
FEP, FEPB, PFA, PFAH, TFE	14	12	21	35	61	79	129	195	299	389	507
	12	9	15	25	44	57	94	142	218	284	370
	10	6	11	18	32	41	68	102	156	203	266
	8	3	6	10	18	23	39	58	89	117	152
	6	2	4	7	13	17	27	41	64	83	108
	4	1	3	5	9	12	19	29	44	58	75
	3	1	2	4	7	10	16	24	37	48	63
	2	1	1	3	6	8	13	20	30	40	52
PFA, PFAH, TFE	1	1	1	2	4	5	9	14	21	28	36

TABLE C7 Maximum Number of Conductors and Fixture Wires in Liquidtight Flexible Metal Conduit (continued)

Type Letters	Conductor Size AWG/kcmil	Conduit Trade Sizes in Inches									
		½	¾	1	1¼	1½	2	2½	3	3½	4
PFA, PFAH, TFE, Z	1/0	1	1	1	3	4	7	11	18	23	30
	2/0	1	1	1	3	4	6	9	14	19	25
	3/0	0	1	1	2	3	5	8	12	16	20
	4/0	0	1	1	1	2	4	6	10	13	17
Z	14	20	26	42	73	95	156	235	360	469	611
	12	14	18	30	52	67	111	167	255	332	434
	10	8	11	18	32	41	68	102	156	203	266
	8	5	7	11	20	26	43	64	99	129	168
	6	4	5	8	14	18	30	45	69	90	118
	4	2	3	5	9	12	20	31	48	62	81
	3	2	2	4	7	9	15	23	35	45	59
	2	1	1	3	6	7	12	19	29	38	49
	1	1	1	2	5	6	10	15	23	30	40
XHH, XHHW, XHHW-2, ZW	14	9	15	25	44	57	93	140	215	280	365
	12	7	12	19	33	43	71	108	165	215	280
	10	5	9	14	25	32	53	80	123	160	209
	8	3	5	8	14	18	29	44	68	89	116
	6	1	3	6	10	13	22	33	50	66	86

	4	1	2	4	7	9	16	24	36	48	62
	3	1	1	3	6	8	13	20	31	40	52
	2	1	1	3	5	7	11	17	26	34	44
XHH, XHHW, XHHW-2	1	1	1	1	4	5	8	12	19	25	33
	1/0	1	1	1	3	4	7	10	16	21	28
	2/0	0	1	1	2	3	6	9	13	17	23
	3/0	0	1	1	1	3	5	7	11	14	19
	4/0	0	1	1	1	2	4	6	9	12	16
	250	0	0	1	1	1	3	5	7	10	13
	300	0	0	1	1	1	3	4	6	8	11
	350	0	0	1	1	1	2	3	5	7	10
	400	0	0	0	1	1	1	3	5	6	8
	500	0	0	0	1	1	1	2	4	5	7
	600	0	0	0	1	1	1	1	3	4	6
	700	0	0	0	1	1	1	1	3	4	5
	750	0	0	0	0	1	1	1	3	3	5
	800	0	0	0	0	1	1	1	2	3	4
	900	0	0	0	0	1	1	1	2	3	4
	1000	0	0	0	0	0	1	1	1	3	3
	1250	0	0	0	0	0	1	1	1	1	3
	1500	0	0	0	0	0	1	1	1	1	2
	1750	0	0	0	0	0	0	1	1	1	2
	2000	0	0	0	0	0	0	1	1	1	2

TABLE C7 Maximum Number of Conductors and Fixture Wires in Liquidtight Flexible Metal Conduit (continued)

Fixture Wires

Type Letters	Conductor Size AWG/kcmil	Conduit Trade Sizes in Inches					
		½	¾	1	1¼	1½	2
FFH-2, RFH-2, RFHH-3	18 16	8 7	15 12	24 20	42 35	54 46	89 75
SF-2, SFF-2	18 16 14	11 9 7	19 15 12	30 25 20	53 44 35	69 57 46	113 93 75
SF-1, SFF-1	18	19	33	53	94	122	199
AF, RFH-1, RFHH-2, TF, TFF, XF, XFF	18	14	24	39	69	90	147
AF, RFHH-2, TF, TFF, XF, XFF	16	11	20	32	56	72	119
AF, XF, XFF	14	9	15	25	44	57	93
TFN, TFFN	18 16	23 17	39 30	63 48	111 85	144 110	236 180

TABLE C7 (continued)

Type Letters	*Conductor Size AWG/kcmil*	*Conduit Trade Sizes in Inches*					
		½	*¾*	*1*	*1¼*	*1½*	*2*
PF, PFF,	18	21	37	60	105	136	223
PGF, PGFF,	16	16	29	46	81	105	173
PAF, PTF,	14	12	21	35	61	79	129
PTFF, PAFF							
HF, HFF,	18	28	48	77	136	176	288
ZF, ZFF,	16	20	35	57	100	129	212
ZHF	14	15	26	42	73	95	156
KF-2,	18	40	70	112	197	255	418
KFF-2	16	28	49	79	139	180	295
	14	19	34	54	95	123	202
	12	13	23	37	65	85	139
	10	9	15	25	44	57	93
KF-1,	18	48	83	134	235	304	499
KFF-1	16	34	58	94	165	214	350
	14	23	39	63	111	144	236
	12	15	26	42	73	95	156
	10	10	17	27	48	62	102
AF, XF, XFF	12	5	8	13	23	30	50
	10	3	6	10	18	23	39

Note: This table is for concentric stranded conductors only. For compact stranded conductors, Table C7A shall be used.

TABLE C7A Maximum Number of Compact Conductors in Liquidtight Flexible Metal Conduit

(Based on Table 1, Chapter 9)

Type Letters	*Conductor Size AWG/kcmil*	*Conduit Trade Sizes in Inches*										
		⅜	*½*	*¾*	*1*	*1¼*	*1½*	*2*	*2½*	*3*	*3½*	*4*
THW, THW-2, THHW	8	1	2	4	7	12	15	25	38	58	76	99
	6	1	1	3	5	9	12	19	29	45	59	77
	4	1	1	2	4	7	9	14	22	34	44	57
	2	1	1	1	3	5	6	11	16	25	32	42
	1	0	1	1	1	3	4	7	11	17	23	30
	1/0	0	1	1	1	3	4	6	10	15	20	26
	2/0	0	0	1	1	2	3	5	8	13	16	21
	3/0	0	0	1	1	1	3	4	7	11	14	18
	4/0	0	0	1	1	1	2	4	6	9	12	15
	250	0	0	0	1	1	1	3	4	7	9	12
	300	0	0	0	1	1	1	2	4	6	8	10
	350	0	0	0	1	1	1	2	3	5	7	9
	400	0	0	0	0	1	1	1	3	5	6	8
	500	0	0	0	0	1	1	1	3	4	5	7
	600	0	0	0	0	1	1	1	1	3	4	6
	700	0	0	0	0	1	1	1	1	3	4	5
	750	0	0	0	0	0	1	1	1	3	3	5
	1000	0	0	0	0	0	1	1	1	1	3	4

THHN, THWN, THWN-2	8	—	—	—	—	—	—	—	—	—	—	—	—
	6	1	2	4	7	13	17	28	43	66	86	112	
	4	1	1	3	4	8	11	17	26	41	53	69	
	2	1	1	1	3	6	7	12	19	29	38	50	
	1	0	1	1	2	4	6	9	14	22	28	37	
	1/0	0	1	1	1	4	5	8	12	19	24	32	
	2/0	0	1	1	1	3	4	6	10	15	20	26	
	3/0	0	0	1	1	2	3	5	8	13	17	22	
	4/0	0	0	1	1	1	3	4	7	10	14	18	
	250	0	0	1	1	1	1	3	5	8	11	14	
	300	0	0	0	1	1	1	3	4	7	9	12	
	350	0	0	0	1	1	1	2	4	6	8	11	
	400	0	0	0	1	1	1	2	3	5	7	9	
	500	0	0	0	0	1	1	1	3	5	6	8	
	600	0	0	0	0	1	1	1	2	4	5	6	
	700	0	0	0	0	1	1	1	1	3	4	6	
	750	0	0	0	0	1	1	1	1	3	4	5	
	1000	0	0	0	0	0	1	1	1	2	3	4	

TABLE C7A Maximum Number of Compact Conductors in Liquidtight Flexible Metal Conduit (continued)

Type Letters	*Conductor Size AWG/kcmil*	*Conduit Trade Sizes in Inches*										
		3/8	*1/2*	*3/4*	*1*	*1¼*	*1½*	*2*	*2½*	*3*	*3½*	*4*
XHHW, XHHW-2	8	1	3	5	9	15	20	33	49	76	98	129
	6	1	2	4	6	11	15	24	37	56	73	95
	4	1	1	3	4	8	11	17	26	41	53	69
	2	1	1	1	3	6	7	12	19	29	38	50
	1	0	1	1	2	4	6	9	14	22	28	37
	1/0	0	1	1	1	4	5	8	12	19	24	32
	2/0	0	1	1	1	3	4	7	10	16	20	27
	3/0	0	0	1	1	2	3	5	8	13	17	22
	4/0	0	0	1	1	1	3	4	7	11	14	18

	250	0	0	1	1	1	1	3	5	8	11	15
	300	0	0	0	1	1	1	3	5	7	9	12
	350	0	0	0	1	1	1	3	4	6	8	11
	400	0	0	0	1	1	1	2	4	6	7	10
	500	0	0	0	0	1	1	1	3	5	6	8
	600	0	0	0	0	1	1	1	2	4	5	6
	700	0	0	0	0	1	1	1	1	3	4	6
	750	0	0	0	0	1	1	1	1	3	4	5
	1000	0	0	0	0	0	1	1	1	2	3	4

Definition: Compact stranding is the result of a manufacturing process where the standard conductor is compressed to the extent that the interstices (voids between strand wires) are virtually eliminated.

TABLE C8 Maximum Number of Conductors and Fixture Wires in Rigid Metal Conduit
(Based on Table 1, Chapter 9)

Type Letters	Conductor Size AWG/kcmil	Conduit Trade Sizes in Inches											
		½	¾	1	1¼	1½	2	2½	3	3½	4	5	6
RH	14	6	10	17	29	39	65	93	143	191	246	387	558
	12	5	8	13	23	32	52	75	115	154	198	311	448
RHH, RHW, RHW-2	14	4	7	12	21	28	46	66	102	136	176	276	398
	12	3	6	10	17	23	38	55	85	113	146	229	330
RH, RHH, RHW, RHW-2	10	3	5	8	14	19	31	44	68	91	118	185	267
	8	1	2	4	7	10	16	23	36	48	61	97	139
	6	1	1	3	6	8	13	18	29	38	49	77	112
	4	1	1	2	4	6	10	14	22	30	38	60	87
	3	1	1	2	4	5	9	12	19	26	34	53	76
	2	1	1	1	3	4	7	11	17	23	29	46	66
	1	0	1	1	1	3	5	7	11	15	19	30	44

	1/0	0	1	1	1	2	4	6	10	13	17	26	38
	2/0	0	1	1	1	2	4	5	8	11	14	23	33
	3/0	0	0	1	1	1	3	4	7	10	12	20	28
	4/0	0	0	1	1	1	3	4	6	8	11	17	24
	250	0	0	0	1	1	1	3	4	6	8	13	18
	300	0	0	0	1	1	1	2	4	5	7	11	16
	350	0	0	0	1	1	1	2	4	5	6	10	15
	400	0	0	0	1	1	1	1	3	4	6	9	13
	500	0	0	0	1	1	1	1	3	4	5	8	11
	600	0	0	0	0	1	1	1	2	3	4	6	9
	700	0	0	0	0	1	1	1	1	3	4	6	8
	750	0	0	0	0	0	1	1	1	3	3	5	8
	800	0	0	0	0	0	1	1	1	2	3	5	7
	900	0	0	0	0	0	1	1	1	2	3	5	7
	1000	0	0	0	0	0	1	1	1	1	3	4	6
	1250	0	0	0	0	0	0	1	1	1	1	3	5
	1500	0	0	0	0	0	0	1	1	1	1	3	4
	1750	0	0	0	0	0	0	1	1	1	1	2	4
	2000	0	0	0	0	0	0	0	1	1	1	2	3

TABLE C8 Maximum Number of Conductors and Fixture Wires in Rigid Metal Conduit (continued)

Type Letters	Conductor Size AWG/kcmil	Conduit Trade Sizes in Inches											
		½	¾	1	1¼	1½	2	2½	3	3½	4	5	6
TW	14	9	15	25	44	59	98	140	216	288	370	581	839
	12	7	12	19	33	45	75	107	165	221	284	446	644
	10	5	9	14	25	34	56	80	123	164	212	332	480
	8	3	5	8	14	19	31	44	68	91	118	185	267
RHH*, RHW*, RHW-2*, THHW, THW, THW-2	14	6	10	17	29	39	65	93	143	191	246	387	558
RHH*, RHW*, RHW-2*, THHW, THW	12	5	8	13	23	32	52	75	115	154	198	311	448
	10	3	6	10	18	25	41	58	90	120	154	242	350
RHH*, RHW*, RHW-2*, THHW, THW, THW-2	8	1	4	6	11	15	24	35	54	72	92	145	209

RHH*, RHW*, RHW-2*, TW, THW, THHW, THW-2	6	1	3	5	8	11	18	27	41	55	71	111	160
	4	1	1	3	6	8	14	20	31	41	53	83	120
	3	1	1	3	5	7	12	17	26	35	45	71	103
	2	1	1	2	4	6	10	14	22	30	38	60	87
	1	1	1	1	3	4	7	10	15	21	27	42	61
	1/0	0	1	1	2	3	6	8	13	18	23	36	52
	2/0	0	1	1	2	3	5	7	11	15	19	31	44
	3/0	0	1	1	1	2	4	6	9	13	16	26	37
	4/0	0	0	1	1	1	3	5	8	10	14	21	31
	250	0	0	1	1	1	3	4	6	8	11	17	25
	300	0	0	1	1	1	2	3	5	7	9	15	22
	350	0	0	0	1	1	1	3	5	6	8	13	19
	400	0	0	0	1	1	1	3	4	6	7	12	17
	500	0	0	0	1	1	1	2	3	5	6	10	14
	600	0	0	0	1	1	1	1	3	4	5	8	12
	700	0	0	0	0	1	1	1	2	3	4	7	10
	750	0	0	0	0	1	1	1	2	3	4	7	10
	800	0	0	0	0	1	1	1	2	3	4	6	9
	900	0	0	0	0	1	1	1	1	3	4	6	8
	1000	0	0	0	0	0	1	1	1	2	3	5	8
	1250	0	0	0	0	0	1	1	1	1	2	4	6
	1500	0	0	0	0	0	1	1	1	1	2	3	5
	1750	0	0	0	0	0	0	1	1	1	1	3	4
	2000	0	0	0	0	0	0	1	1	1	1	3	4

* Types RHH, RHW, and RHW-2 without outer covering.

TABLE C8 Maximum Number of Conductors and Fixture Wires in Rigid Metal Conduit (continued)

Type Letters	*Conductor Size AWG/kcmil*	*Conduit Trade Sizes in Inches*											
		½	¾	1	1¼	1½	2	2½	3	3½	4	5	6
THHN, THWN, THWN-2	14	13	22	36	63	85	140	200	309	412	531	833	1202
	12	9	16	26	46	62	102	146	225	301	387	608	877
	10	6	10	17	29	39	64	92	142	189	244	383	552
	8	3	6	9	16	22	37	53	82	109	140	221	318
	6	2	4	7	12	16	27	38	59	79	101	159	230
	4	1	2	4	7	10	16	23	36	48	62	98	141
	3	1	1	3	6	8	14	20	31	41	53	83	120
	2	1	1	3	5	7	11	17	26	34	44	70	100
	1	1	1	1	4	5	8	12	19	25	33	51	74
	1/0	1	1	1	3	4	7	10	16	21	27	43	63
	2/0	0	1	1	2	3	6	8	13	18	23	36	52
	3/0	0	1	1	1	3	5	7	11	15	19	30	43
	4/0	0	1	1	1	2	4	6	9	12	16	25	36
	250	0	0	1	1	1	3	5	7	10	13	20	29
	300	0	0	1	1	1	3	4	6	8	11	17	25
	350	0	0	1	1	1	2	3	5	7	10	15	22

	400	0	0	1	1	1	2	3	5	7	8	13	20
	500	0	0	0	1	1	1	2	4	5	7	11	16
	600	0	0	0	1	1	1	1	3	4	6	9	13
	700	0	0	0	1	1	1	1	3	4	5	8	11
	750	0	0	0	0	1	1	1	3	4	5	7	11
	800	0	0	0	0	1	1	1	2	3	4	7	10
	900	0	0	0	0	1	1	1	2	3	4	6	9
	1000	0	0	0	0	1	1	1	1	3	4	6	8
FEP, FEPB, PFA, PFAH, TFE	14	12	22	35	61	83	136	194	300	400	515	808	1166
	12	9	16	26	44	60	99	142	219	292	376	590	851
	10	6	11	18	32	43	71	102	157	209	269	423	610
	8	3	6	10	18	25	41	58	90	120	154	242	350
	6	2	4	7	13	17	29	41	64	85	110	172	249
	4	1	3	5	9	12	20	29	44	59	77	120	174
	3	1	2	4	7	10	17	24	37	50	64	100	145
	2	1	1	3	6	8	14	20	31	41	53	83	120
PFA, PFAH, TFE	1	1	1	2	4	6	9	14	21	28	37	57	83

TABLE C8 Maximum Number of Conductors and Fixture Wires in Rigid Metal Conduit (continued)

Type Letters	Conductor Size AWG/kcmil	Conduit Trade Sizes in Inches											
		½	¾	1	1¼	1½	2	2½	3	3½	4	5	6
PFA, PFAH, TFE, Z	1/0	1	1	1	3	5	8	11	18	24	30	48	69
	2/0	1	1	1	3	4	6	9	14	19	25	40	57
	3/0	0	1	1	2	3	5	8	12	16	21	33	47
	4/0	0	1	1	1	2	4	6	10	13	17	27	39
Z	14	15	26	42	73	100	164	234	361	482	621	974	1405
	12	10	18	30	52	71	116	166	256	342	440	691	997
	10	6	11	18	32	43	71	102	157	209	269	423	610
	8	4	7	11	20	27	45	64	99	132	170	267	386
	6	3	5	8	14	19	31	45	69	93	120	188	271
	4	1	3	5	9	13	22	31	48	64	82	129	186
	3	1	2	4	7	9	16	22	35	47	60	94	136
	2	1	1	3	6	8	13	19	29	39	50	78	113
	1	1	1	2	5	6	10	15	23	31	40	63	92
XHH, XHHW, XHHW-2, ZW	14	9	15	25	44	59	98	140	216	288	370	581	839
	12	7	12	19	33	45	75	107	165	221	284	446	644
	10	5	9	14	25	34	56	80	123	164	212	332	480
	8	3	5	8	14	19	31	44	68	91	118	185	267

	6	1	3	6	10	14	23	33	51	68	87	137	197
	4	1	2	4	7	10	16	24	37	49	63	99	143
	3	1	1	3	6	8	14	20	31	41	53	84	121
	2	1	1	3	5	7	12	17	26	35	45	70	101
XHH, XHHW, XHHW-2	1	1	1	1	4	5	9	12	19	26	33	52	76
	1/0	1	1	1	3	4	7	10	16	22	28	44	64
	2/0	0	1	1	2	3	6	9	13	18	23	37	53
	3/0	0	1	1	1	3	5	7	11	15	19	30	44
	4/0	0	1	1	1	2	4	6	9	12	16	25	36
	250	0	0	1	1	1	3	5	7	10	13	20	30
	300	0	0	1	1	1	3	4	6	9	11	18	25
	350	0	0	1	1	1	2	3	6	7	10	15	22
	400	0	0	1	1	1	2	3	5	7	9	14	20
	500	0	0	0	1	1	1	2	4	5	7	11	16
	600	0	0	0	1	1	1	1	3	4	6	9	13
	700	0	0	0	1	1	1	1	3	4	5	8	11
	750	0	0	0	0	1	1	1	3	4	5	7	[illegible]
	800	0	0	0	0	1	1	1	2	3	4	7	[illegible]
	900	0	0	0	0	1	1	1	2	3	4	6	[illegible]
	1000	0	0	0	0	1	1	1	1	3	4	6	[illegible]
	1250	0	0	0	0	0	1	1	1	2	3	4	[illegible]
	1500	0	0	0	0	0	1	1	1	1	2	4	[illegible]
	1750	0	0	0	0	0	0	1	1	1	1	3	[illegible]
	2000	0	0	0	0	0	0	1	1	1	1	3	[illegible]

TABLE C8 Maximum Number of Conductors and Fixture Wires in Rigid Metal Conduit (continued)

Fixture Wires

Type Letters	Conductor Size AWG/kcmil	Conduit Trade Sizes in Inches					
		½	¾	1	1¼	1½	2
FFH-2, RFH-2, RFHH-3	18 16	8 7	15 12	24 20	42 35	57 48	94 79
SF-2, SFF-2	18 16 14	11 9 7	19 15 12	31 25 20	53 44 35	72 59 48	118 98 79
SF-1, SFF-1	18	19	33	54	94	127	209
AF, RFH-1, RFHH-2, TF, TFF, XF, XFF	18	14	25	40	69	94	155
AF, RFHH-2, TF, TFF, XF, XFF	16	11	20	32	56	76	125
AF, XF, XFF	14	9	15	25	44	59	98
TFN, TFFN	18 16	23 17	40 30	64 49	111 84	150 115	248 189
PF, PFF, PGF, PGFF, PAF, PTF, PTFF, PAFF	18 16 14	21 16 12	38 29 22	61 47 35	105 81 61	143 110 83	235 181 136

TABLE C8 (continued)

Type Letters	Conductor Size AWG/kcmil	Conduit Trade Sizes in Inches					
		½	¾	1	1¼	1½	2
HF, HFF,	18	28	48	79	135	184	303
ZF, ZFF,	16	20	36	58	100	136	223
ZHF	14	15	26	42	73	100	164
KF-2,	18	40	71	114	197	267	439
KFF-2	16	28	50	80	138	188	310
	14	19	34	55	95	129	213
	12	13	23	38	65	89	146
	10	9	15	25	44	59	98
KF-1,	18	48	84	136	235	318	524
KFF-1	16	34	59	96	165	224	368
	14	23	40	64	111	150	248
	12	15	26	42	73	100	164
	10	10	17	28	48	65	107
AF, XF,	12	5	8	13	23	32	52
XFF	10	3	6	10	18	25	41

Note: This table is for concentric stranded conductors only. For compact stranded conductors, Table C8A shall be used.

TABLE C8A Maximum Number of Compact Conductors in Rigid Metal Conduit
(Based on Table 1, Chapter 9)

Type Letters	*Conductor Size AWG/kcmil*	*Conduit Trade Sizes in Inches*											
		½	*¾*	*1*	*1¼*	*1½*	*2*	*2½*	*3*	*3½*	*4*	*5*	*6*
THW, THW-2, THHW	8	2	4	7	12	16	26	38	59	78	101	158	228
	6	1	3	5	9	12	20	29	45	60	78	122	176
	4	1	2	4	7	9	15	22	34	45	58	91	132
	2	1	1	3	5	7	11	16	25	33	43	67	97
	1	1	1	1	3	5	8	11	17	23	30	47	68
	1/0	1	1	1	3	4	7	10	15	20	26	41	59
	2/0	0	1	1	2	3	6	8	13	17	22	34	50
	3/0	0	1	1	1	3	5	7	11	14	19	29	42
	4/0	0	1	1	1	2	4	6	9	12	15	24	35
	250	0	0	1	1	1	3	4	7	9	12	19	28
	300	0	0	1	1	1	3	4	6	8	11	17	24
	350	0	0	1	1	1	2	3	5	7	9	15	22
	400	0	0	1	1	1	1	3	5	7	8	13	20
	500	0	0	0	1	1	1	3	4	5	7	11	17
	600	0	0	0	1	1	1	1	3	4	6	9	13
	700	0	0	0	1	1	1	1	3	4	5	8	12
	750	0	0	0	0	1	1	1	3	4	5	7	11
	1000	0	0	0	0	1	1	1	1	3	4	6	9

THHN, THWN, THWN-2	8	—	—	—	—	—	—	—	—	—	—	—	—
	6	2	5	8	13	18	30	43	66	88	114	179	258
	4	1	3	5	8	11	18	26	41	55	70	110	159
	2	1	1	3	6	8	13	19	29	39	50	79	114
	1	1	1	2	4	6	10	14	22	29	38	60	86
	1/0	1	1	1	4	5	8	12	19	25	32	51	73
	2/0	1	1	1	3	4	7	10	15	21	26	42	60
	3/0	0	1	1	2	3	6	8	13	17	22	35	51
	4/0	0	1	1	1	3	5	7	10	14	18	29	42
	250	0	1	1	1	2	4	5	8	11	14	23	33
	300	0	0	1	1	1	3	4	7	10	12	20	28
	350	0	0	2	1	1	3	4	6	8	11	17	25
	400	0	0	1	1	1	2	3	5	7	10	15	22
	500	0	0	0	1	1	1	3	5	6	8	13	19
	600	0	0	0	1	1	1	2	4	5	6	10	15
	700	0	0	0	1	1	1	1	3	4	6	9	13
	750	0	0	0	1	1	1	1	3	4	5	9	13
	1000	0	0	0	0	1	1	1	2	3	4	6	9

TABLE C8A Maximum Number of Compact Conductors in Rigid Metal Conduit (continued)

Type Letters	*Conductor Size AWG/kcmil*	*Conduit Trade Sizes in Inches*											
		½	¾	*1*	*1¼*	*1½*	*2*	*2½*	*3*	*3½*	*4*	*5*	*6*
XHHW, XHHW-2	8	3	5	9	15	21	34	49	76	101	130	205	296
	6	2	4	6	11	15	25	36	56	75	97	152	220
	4	1	3	5	8	11	18	26	41	55	70	110	159
	2	1	1	3	6	8	13	19	29	39	50	79	114
	1	1	1	2	4	6	10	14	22	29	38	60	86
	1/0	1	1	1	4	5	8	12	19	25	32	51	73
	2/0	1	1	1	3	4	7	10	16	21	27	43	62
	3/0	0	1	1	2	3	6	8	13	17	22	35	51
	4/0	0	1	1	1	3	5	7	11	14	19	29	42
	250	0	1	1	1	2	4	5	8	11	15	23	34

	300	0	0	1	1	1	3	5	7	10	13	20	29
	350	0	0	1	1	1	3	4	6	9	11	18	25
	400	0	0	1	1	1	2	4	6	8	10	16	23
	500	0	0	0	1	1	1	3	5	6	8	13	19
	600	0	0	0	1	1	1	2	4	5	7	10	15
	700	0	0	0	1	1	1	1	3	4	6	9	13
	750	0	0	0	1	1	1	1	3	4	5	8	12
	1000	0	0	0	0	1	1	1	2	3	4	7	10

Definition: Compact stranding is the result of a manufacturing process where the standard conductor is compressed to the extent that the interstices (voids between strand wires) are virtually eliminated.

TABLE C9 Maximum Number of Conductors and Fixture Wires in Rigid PVC Conduit, Schedule 80
(Based on Table 1, Chapter 9)

Type Letters	*Conductor Size AWG/kcmil*	*Conduit Trade Sizes in Inches*											
		½	*¾*	*1*	*1¼*	*1½*	*2*	*2½*	*3*	*3½*	*4*	*5*	*6*
RH	14	4	8	13	23	32	55	79	123	166	215	341	490
	12	3	6	10	19	26	44	63	99	133	173	274	394
RHH, RHW,	14	3	5	9	17	23	39	56	88	118	153	243	349
RHW-2	12	2	4	7	14	19	32	46	73	98	127	202	290
RH, RHH,	10	1	3	6	11	15	26	37	59	79	103	163	234
RHW,	8	1	1	3	6	8	13	19	31	41	54	85	122
RHW-2	6	1	1	2	4	6	11	16	24	33	43	68	98
	4	1	1	1	3	5	8	12	19	26	33	53	77
	3	0	1	1	3	4	7	11	17	23	29	47	67
	2	0	1	1	3	4	6	9	14	20	25	41	58
	1	0	1	1	1	2	4	6	9	13	17	27	38

1/0	0	0	1	1	1	3	5	8	11	15	23	33
2/0	0	0	1	1	1	3	4	7	10	13	20	29
3/0	0	0	1	1	1	3	4	6	8	11	17	25
4/0	0	0	0	1	1	2	3	5	7	9	15	21
250	0	0	0	1	1	1	2	4	5	7	11	16
300	0	0	0	1	1	1	2	3	5	6	10	14
350	0	0	0	1	1	1	1	3	4	5	9	13
400	0	0	0	0	1	1	1	3	4	5	8	12
500	0	0	0	0	1	1	1	2	3	4	7	10
600	0	0	0	0	0	1	1	1	3	3	6	8
700	0	0	0	0	0	1	1	1	2	3	5	7
750	0	0	0	0	0	1	1	1	2	3	5	7
800	0	0	0	0	0	1	1	1	2	3	4	7
1000	0	0	0	0	0	1	1	1	1	2	4	5
1250	0	0	0	0	0	0	1	1	1	1	3	4
1500	0	0	0	0	0	0	1	1	1	1	2	4
1750	0	0	0	0	0	0	0	1	1	1	2	3
2000	0	0	0	0	0	0	0	1	1	1	1	3

TABLE C9 Maximum Number of Conductors and Fixture Wires in Rigid PVC Conduit, Schedule 80 (continued)

Type Letters	*Conductor Size AWG/kcmil*	*Conduit Trade Sizes in Inches*											
		½	*¾*	*1*	*1¼*	*1½*	*2*	*2½*	*3*	*3½*	*4*	*5*	*6*
TW	14	6	11	20	35	49	82	118	185	250	324	514	736
	12	5	9	15	27	38	63	91	142	192	248	394	565
	10	3	6	11	20	28	47	67	106	143	185	294	421
	8	1	3	6	11	15	26	37	59	79	103	163	234
RHH*, RHW*, RHW-2*, THHW, THW, THW-2	14	4	8	13	23	32	55	79	123	166	215	341	490
RHH*, RHW*, RHW-2*, THHW, THW	12	3	6	10	19	26	44	63	99	133	173	274	394
	10	2	5	8	15	20	34	49	77	104	135	214	307
RHH*, RHW*, RHW-2*, THHW, THW, THW-2	8	1	3	5	9	12	20	29	46	62	81	128	184

RHH*, RHW*, RHW-2*, TW, THW, THHW, THW-2	6	1	1	3	7	9	16	22	35	48	62	98	141
	4	1	1	3	5	7	12	17	26	35	46	73	105
	3	1	1	2	4	6	10	14	22	30	39	63	90
	2	1	1	1	3	5	8	12	19	26	33	53	77
	1	0	1	1	2	3	6	8	13	18	23	37	54
	1/0	0	1	1	1	3	5	7	11	15	20	32	46
	2/0	0	1	1	1	2	4	6	10	13	17	27	39
	3/0	0	0	1	1	1	3	5	8	11	14	23	33
	4/0	0	0	1	1	1	3	4	7	9	12	19	27
	250	0	0	0	1	1	2	3	5	7	9	15	22
	300	0	0	0	1	1	1	3	5	6	8	13	19
	350	0	0	0	1	1	1	2	4	6	7	12	17
	400	0	0	0	1	1	1	2	4	5	7	10	15
	500	0	0	0	1	1	1	1	3	4	5	9	13
	600	0	0	0	0	1	1	1	2	3	4	7	10
	700	0	0	0	0	1	1	1	2	3	4	6	9
	750	0	0	0	0	0	1	1	1	3	4	6	8
	800	0	0	0	0	0	1	1	1	3	3	6	8
	900	0	0	0	0	0	1	1	1	2	3	5	7
	1000	0	0	0	0	0	1	1	1	2	3	5	7
	1250	0	0	0	0	0	1	1	1	1	2	4	5
	1500	0	0	0	0	0	0	1	1	1	1	3	4
	1750	0	0	0	0	0	0	1	1	1	1	3	4
	2000	0	0	0	0	0	0	0	1	1	1	2	3

*Types RHH, RHW, and RHW-2 without outer covering.

TABLE C9 Maximum Number of Conductors and Fixture Wires in Rigid PVC Conduit, Schedule 80 (continued)

Type Letters	*Conductor Size AWG/kcmil*	*Conduit Trade Sizes in Inches*											
		½	*¾*	*1*	*1¼*	*1½*	*2*	*2½*	*3*	*3½*	*4*	*5*	*6*
THHN,	14	9	17	28	51	70	118	170	265	358	464	736	1055
THWN,	12	6	12	20	37	51	86	124	193	261	338	537	770
THWN-2	10	4	7	13	23	32	54	78	122	164	213	338	485
	8	2	4	7	13	18	31	45	70	95	123	195	279
	6	1	3	5	9	13	22	32	51	68	89	141	202
	4	1	1	3	6	8	14	20	31	42	54	86	124
	3	1	1	3	5	7	12	17	26	35	46	73	105
	2	1	1	2	4	6	10	14	22	30	39	61	88
	1	0	1	1	3	4	7	10	16	22	29	45	65
	1/0	0	1	1	2	3	6	9	14	18	24	38	55
	2/0	0	1	1	1	3	5	7	11	15	20	32	46
	3/0	0	1	1	1	2	4	6	9	13	17	26	38
	4/0	0	0	1	1	1	3	5	8	10	14	22	31
	250	0	0	1	1	1	3	4	6	8	11	18	25
	300	0	0	0	1	1	2	3	5	7	9	15	22
	350	0	0	0	1	1	1	3	5	6	8	13	19
	400	0	0	0	1	1	1	3	4	6	7	12	17
	500	0	0	0	1	1	1	2	3	5	6	10	14
	600	0	0	0	0	1	1	1	3	4	5	8	12

	700	0	0	0	0	1	1	1	2	3	4	7	10
	750	0	0	0	0	1	1	1	2	3	4	7	9
	800	0	0	0	0	1	1	1	2	3	4	6	9
	900	0	0	0	0	0	1	1	1	3	3	6	8
	1000	0	0	0	0	0	1	1	1	2	3	5	7
FEP, FEPB, PFA, PFAH, TFE	14	8	16	27	49	68	115	164	257	347	450	714	1024
	12	6	12	20	36	50	84	120	188	253	328	521	747
	10	4	8	14	26	36	60	86	135	182	235	374	536
	8	2	5	8	15	20	34	49	77	104	135	214	307
	6	1	3	6	10	14	24	35	55	74	96	152	218
	4	1	2	4	7	10	17	24	38	52	67	106	153
	3	1	1	3	6	8	14	20	32	43	56	89	127
	2	1	1	3	5	7	12	17	26	35	46	73	105
PFA, PFAH, TFE	1	1	1	1	3	5	8	11	18	25	32	51	73
PFA, PFAH, TFE, Z	1/0	0	1	1	3	4	7	10	15	20	27	42	61
	2/0	0	1	1	2	3	5	8	12	17	22	35	50
	3/0	0	1	1	1	2	4	6	10	14	18	29	41
	4/0	0	0	1	1	1	4	5	8	11	15	24	34

TABLE C9 Maximum Number of Conductors and Fixture Wires in Rigid PVC Conduit, Schedule 80 (continued)

Type Letters	*Conductor Size AWG/kcmil*	*Conduit Trade Sizes in Inches*											
		½	*¾*	*1*	*1¼*	*1½*	*2*	*2½*	*3*	*3½*	*4*	*5*	*6*
Z	14	10	19	33	59	82	138	198	310	418	542	860	1233
	12	7	14	23	42	58	98	141	220	297	385	610	875
	10	4	8	14	26	36	60	86	135	182	235	374	536
	8	3	5	9	16	22	38	54	85	115	149	236	339
	6	2	4	6	11	16	26	38	60	81	104	166	238
	4	1	2	4	8	11	18	26	41	55	72	114	164
	3	1	2	3	5	8	13	19	30	40	52	83	119
	2	1	1	2	5	6	11	16	25	33	43	69	99
	1	0	1	2	4	5	9	13	20	27	35	56	80
XHH, XHHW, XHHW-2, ZW	14	6	11	20	35	49	82	118	185	250	324	514	736
	12	5	9	15	27	38	63	91	142	192	248	394	565
	10	3	6	11	20	28	47	67	106	143	185	294	421
	8	1	3	6	11	15	26	37	59	79	103	163	234
	6	1	2	4	8	11	19	28	43	59	76	121	173
	4	1	1	3	6	8	14	20	31	42	55	87	125
	3	1	1	3	5	7	12	17	26	36	47	74	106
	2	1	1	2	4	6	10	14	22	30	39	62	89

XHH, XHHW, XHHW-2	1	0	1	1	3	4	7	10	16	22	29	46	66
	1/0	0	1	1	2	3	6	9	14	19	24	39	56
	2/0	0	1	1	1	3	5	7	11	16	20	32	46
	3/0	0	1	1	1	2	4	6	9	13	17	27	38
	4/0	0	0	1	1	1	3	5	8	11	14	22	32
	250	0	0	1	1	1	3	4	6	9	11	18	26
	300	0	0	1	1	1	2	3	5	7	10	15	22
	350	0	0	0	1	1	1	3	5	6	8	14	20
	400	0	0	0	1	1	1	3	4	6	7	12	17
	500	0	0	0	1	1	1	2	3	5	6	10	14
	600	0	0	0	0	1	1	1	3	4	5	8	11
	700	0	0	0	0	1	1	1	2	3	4	7	10
	750	0	0	0	0	1	1	1	2	3	4	6	9
	800	0	0	0	0	1	1	1	1	3	4	6	9
	900	0	0	0	0	0	1	1		3	3	5	8
	1000	0	0	0	0	0	1	1	1	2	3	5	7
	1250	0	0	0	0	0	1	1	1	1	2	4	6
	1500	0	0	0	0	0	0	1	1	1	1	3	5
	1750	0	0	0	0	0	0	1	1	1	1	3	4
	2000	0	0	0	0	0	0	1	1	1	1	2	4

TABLE C9 Maximum Number of Conductors and Fixture Wires in Rigid PVC Conduit, Schedule 80 (continued)

Fixture Wires

Type Letters	Conductor Size AWG/kcmil	Conduit Trade Sizes in Inches					
		½	¾	1	1¼	1½	2
FFH-2, RFH-2, RFHH-3	18 16	6 5	11 9	19 16	34 28	47 39	79 67
SF-2, SFF-2	18 16 14	7 6 5	14 11 9	24 20 16	43 35 28	59 49 39	100 82 67
SF-1, SFF-1	18	13	25	42	76	105	177
AF, RFH-1, RFHH-2, TF, TFF, XF, XFF	18	10	18	31	56	77	130
AF, RFHH-2, TF, TFF, XF, XFF	16	8	15	25	45	62	105
AF, XF, XFF	14	6	11	20	35	49	82
TFN, TFFN	18 16	16 12	29 22	50 38	90 68	124 95	209 159
PF, PFF, PGF, PGFF, PAF, PTF, PTFF, PAFF	18 16 14	15 11 8	28 22 16	47 36 27	85 66 49	118 91 68	198 153 115

TABLE C9 (continued)

Type Letters	*Conductor Size AWG/kcmil*	*Conduit Trade Sizes in Inches*					
		½	*¾*	*1*	*1¼*	*1½*	*2*
ZF, ZFF,	18	19	36	61	110	152	255
ZHF, HF,	16	14	27	45	81	112	188
HFF	14	10	19	33	59	82	138
KF-2,	18	28	53	88	159	220	371
KFF-2	16	19	37	62	112	155	261
	14	13	25	43	77	107	179
	12	9	17	29	53	73	123
	10	6	11	20	35	49	82
KF-1,	18	33	63	106	190	263	442
KFF-1	16	23	44	74	133	185	310
	14	16	29	50	90	124	209
	12	10	19	33	59	82	138
	10	7	13	21	39	54	90
AF, XF,	12	3	6	10	19	26	44
XFF	10	2	5	8	15	20	34

Note: This table is for concentric stranded conductors only. For compact stranded conductors, Table C9A shall be used.

TABLE C9A Maximum Number of Compact Conductors in Rigid PVC Conduit, Schedule 80
(Based on Table 1, Chapter 9)

Type Letters	Conductor Size AWG/kcmil	Conduit Trade Sizes in Inches											
		½	¾	1	1¼	1½	2	2½	3	3½	4	5	6
THW, THW-2, THWW	8	1	3	5	9	13	22	32	50	68	88	140	200
	6	1	2	4	7	10	17	25	39	52	68	108	155
	4	1	1	3	5	7	13	18	29	39	51	81	116
	2	1	1	1	4	5	9	13	21	29	37	60	85
	1	0	1	1	3	4	6	9	15	20	26	42	60
	1/0	0	1	1	2	3	6	8	13	17	23	36	52
	2/0	0	1	1	1	3	5	7	11	15	19	30	44
	3/0	0	0	1	1	2	4	6	9	12	16	26	37
	4/0	0	0	1	1	1	3	5	8	10	13	22	31
	250	0	0	1	1	1	2	4	6	8	11	17	25
	300	0	0	0	1	1	2	3	5	7	9	15	21
	350	0	0	0	1	1	1	3	5	6	8	13	19
	400	0	0	0	1	1	1	3	4	6	7	12	17
	500	0	0	0	1	1	1	2	3	5	6	10	14
	600	0	0	0	0	1	1	1	3	4	5	8	12
	700	0	0	0	0	1	1	1	2	3	4	7	10
	750	0	0	0	0	1	1	1	2	3	4	7	10
	1000	0	0	0	0	0	1	1	1	2	3	5	8

THHN,	8	—	—	—	—	—	—	—	—	—	—	—	—
THWN,	6	1	3	6	11	15	25	36	57	77	99	158	226
THWN-2	4	1	1	3	6	9	15	22	35	47	61	98	140
	2	1	1	2	5	6	11	16	25	34	44	70	100
	1	1	1	1	3	5	8	12	19	25	33	53	75
	1/0	0	1	1	3	4	7	10	16	22	28	45	64
	2/0	0	1	1	2	3	6	8	13	18	23	37	53
	3/0	0	1	1	1	3	5	7	11	15	19	31	44
	4/0	0	0	1	1	2	4	6	9	12	16	25	37
	250	0	0	1	1	1	3	4	7	10	12	20	29
	300	0	0	1	1	1	3	4	6	8	11	17	25
	350	0	0	0	1	1	2	3	5	7	9	15	22
	400	0	0	0	1	1	1	3	5	6	8	13	19
	500	0	0	0	1	1	1	2	4	5	7	11	16
	600	0	0	0	1	1	1	1	3	4	6	9	13
	700	0	0	0	0	1	1	1	3	4	5	8	12
	750	0	0	0	0	1	1	1	3	4	5	8	11
	1000	0	0	0	0	0	1	1	1	3	3	5	8

TABLE C9A Maximum Number of Compact Conductors in Rigid PVC Conduit, Schedule 80 (continued)

Type Letters	*Conductor Size AWG/kcmil*	*Conduit Trade Sizes in Inches*											
		½	*¾*	*1*	*1¼*	*1½*	*2*	*2½*	*3*	*3½*	*4*	*5*	*6*
XHHW, XHHW-2	8	1	4	7	12	17	29	42	65	88	114	181	260
	6	1	3	5	9	13	21	31	48	65	85	134	193
	4	1	1	3	6	9	15	22	35	47	61	98	140
	2	1	1	2	5	6	11	16	25	34	44	70	100
	1	1	1	1	3	5	8	12	19	25	33	53	75
	1/0	0	1	1	3	4	7	10	16	22	28	45	64
	2/0	0	1	1	2	3	6	8	13	18	24	38	54
	3/0	0	1	1	1	3	5	7	11	15	19	31	44
	4/0	0	0	1	1	2	4	6	9	12	16	26	37
	250	0	0	1	1	1	3	5	7	10	13	21	30

	Size	Col 1	Col 2	Col 3	Col 4	Col 5	Col 6	Col 7	Col 8	Col 9	Col 10	Col 11	Col 12
	300	0	0	1	1	1	3	4	6	8	11	17	25
	350	0	0	1	1	1	2	3	5	7	10	15	22
	400	0	0	0	1	1	1	3	5	7	9	14	20
	500	0	0	0	1	1	1	2	4	5	7	11	17
	600	0	0	0	1	1	1	1	3	4	6	9	13
	700	0	0	0	0	1	1	1	3	4	5	8	12
	750	0	0	0	0	1	1	1	2	3	5	7	11
	1000	0	0	0	0	0	1	1	1	3	3	6	8

Definition: Compact stranding is the result of a manufacturing process where the standard conductor is compressed to the extent that the interstices (voids between strand wires) are virtually eliminated.

TABLE C10 Maximum Number of Conductors and Fixture Wires in Rigid PVC Conduit, Schedule 40 and HDPE Conduit

(Based on Table 1, Chapter 9)

Type Letters	*Conductor Size AWG/kcmil*	*Conduit Trade Sizes in Inches*											
		½	*¾*	*1*	*1¼*	*1½*	*2*	*2½*	*3*	*3½*	*4*	*5*	*6*
RH	14	5	9	16	28	38	63	90	139	186	240	378	546
	12	4	8	12	22	30	50	72	112	150	193	304	439
RHH, RHW, RHW-2	14	4	7	11	20	27	45	64	99	133	171	269	390
	12	3	5	9	16	22	37	53	82	110	142	224	323
RH, RHH, RHW, RHW-2	10	2	4	7	13	18	30	43	66	89	115	181	261
	8	1	2	4	7	9	15	22	35	46	60	94	137
	6	1	1	3	5	7	12	18	28	37	48	76	109
	4	1	1	2	4	6	10	14	22	29	37	59	85
	3	1	1	1	4	5	8	12	19	25	33	52	75
	2	1	1	1	3	4	7	10	16	22	28	45	65
	1	0	1	1	1	3	5	7	11	14	19	29	43

	1/0	0	1	1	1	2	4	6	9	13	16	26	37
	2/0	0	0	1	1	1	3	5	8	11	14	22	32
	3/0	0	0	1	1	1	3	4	7	9	12	19	28
	4/0	0	0	1	1	1	2	4	6	8	10	16	24
	250	0	0	0	1	1	1	3	4	6	8	12	18
	300	0	0	0	1	1	1	2	4	5	7	11	16
	350	0	0	0	1	1	1	2	3	5	6	10	14
	400	0	0	0	1	1	1	1	3	4	6	9	13
	500	0	0	0	0	1	1	1	3	4	5	8	11
	600	0	0	0	0	1	1	1	2	3	4	6	9
	700	0	0	0	0	0	1	1	1	3	3	6	8
	750	0	0	0	0	0	1	1	1	2	3	5	8
	800	0	0	0	0	0	1	1	1	2	3	5	7
	900	0	0	0	0	0	1	1	1	2	3	5	7
	1000	0	0	0	0	0	1	1	1	1	3	4	6
	1250	0	0	0	0	0	0	1	1	1	1	3	5
	1500	0	0	0	0	0	0	1	1	1	1	3	4
	1750	0	0	0	0	0	0	1	1	1	1	2	3
	2000	0	0	0	0	0	0	0	1	1	1	2	3

TABLE C10 Maximum Number of Conductors and Fixture Wires in Rigid PVC Conduit, Schedule 40 and HDPE Conduit (continued)

(Based on Table 1, Chapter 9)

Type Letters	*Conductor Size AWG/kcmil*	*Conduit Trade Sizes in Inches*											
		½	*¾*	*1*	*1¼*	*1½*	*2*	*2½*	*3*	*3½*	*4*	*5*	*6*
TW	14	8	14	24	42	57	94	135	209	280	361	568	822
	12	6	11	18	32	44	72	103	160	215	277	436	631
	10	4	8	13	24	32	54	77	119	160	206	325	470
	8	2	4	7	13	18	30	43	66	89	115	181	261
RHH*, RHW*, RHW-2*, THHW, THW, THW-2	14	5	9	16	28	38	63	90	139	186	240	378	546
RHH*, RHW*, RHW-2*, THHW, THW*	12	4	8	12	22	30	50	72	112	150	193	304	439
	10	3	6	10	17	24	39	56	87	117	150	247	343
RHH*, RHW*, RHW-2*, THHW, THW, THW-2	8	1	3	6	10	14	23	33	52	70	90	142	205

RHH*, RHW*,	6	1	2	4	8	11	18	26	40	53	69	109	157
RHW-2*,	4	1	1	3	6	8	13	19	30	40	51	81	117
TW, THW,	3	1	1	3	5	7	11	16	25	34	44	69	100
THHW,	2	1	1	2	4	6	10	14	22	29	37	59	85
THW-2	1	0	1	1	3	4	7	10	15	20	26	41	60
	1/0	0	1	1	2	3	6	8	13	17	22	35	51
	2/0	0	1	1	1	3	5	7	11	15	19	30	43
	3/0	0	1	1	1	2	4	6	9	12	16	25	36
	4/0	0	0	1	1	1	3	5	8	10	13	21	30
	250	0	0	1	1	1	3	4	6	8	11	17	25
	300	0	0	1	1	1	2	3	5	7	9	15	21
	350	0	0	0	1	1	1	3	5	6	8	13	19
	400	0	0	0	1	1	1	3	4	6	7	12	17
	500	0	0	0	1	1	1	2	3	5	6	10	14
	600	0	0	0	0	1	1	1	3	4	5	8	11
	700	0	0	0	0	1	1	1	2	3	4	7	10
	750	0	0	0	0	1	1	1	2	3	4	6	10
	800	0	0	0	0	1	1	1	2	3	4	6	9
	900	0	0	0	0	0	1	1	1	3	3	6	8
	1000	0	0	0	0	0	1	1	1	2	3	5	7
	1250	0	0	0	0	0	1	1	1	1	2	4	6
	1500	0	0	0	0	0	1	1	1	1	1	3	5
	1750	0	0	0	0	0	0	1	1	1	1	3	4
	2000	0	0	0	0	0	0	1	1	1	1	3	4

* Types RHH, RHW, and RHW-2 without outer covering.

TABLE C10 Maximum Number of Conductors and Fixture in Rigid PVC Conduit, Schedule 40 and HDPE Conduit (continued)

Type Letters	*Conductor Size AWG/kcmil*	*Conduit Trade Sizes in Inches*											
		½	*¾*	*1*	*1¼*	*1½*	*2*	*2½*	*3*	*3½*	*4*	*5*	*6*
THHN, THWN, THWN-2	14	11	21	34	60	82	135	193	299	401	517	815	1178
	12	8	15	25	43	59	99	141	218	293	377	594	859
	10	5	9	15	27	37	62	89	137	184	238	374	541
	8	3	5	9	16	21	36	51	79	106	137	216	312
	6	1	4	6	11	15	26	37	57	77	99	156	225
	4	1	2	4	7	9	16	22	35	47	61	96	138
	3	1	1	3	6	8	13	19	30	40	51	81	117
	2	1	1	3	5	7	11	16	25	33	43	68	98
	1	1	1	1	3	5	8	12	18	25	32	50	73
	1/0	1	1	1	3	4	7	10	15	21	27	42	61
	2/0	0	1	1	2	3	6	8	13	17	22	35	51
	3/0	0	1	1	1	3	5	7	11	14	18	29	42
	4/0	0	1	1	1	2	4	6	9	12	15	24	35
	250	0	0	1	1	1	3	4	7	10	12	20	28
	300	0	0	1	1	1	3	4	6	8	11	17	24

	350	0	0	1	1	1	2	3	5	7	9	15	21
	400	0	0	0	1	1	1	3	5	6	8	13	19
	500	0	0	0	1	1	1	2	4	5	7	11	16
	600	0	0	0	1	1	1	1	3	4	5	9	13
	700	0	0	0	0	1	1	1	3	4	5	8	11
	750	0	0	0	0	1	1	1	2	3	4	7	11
	800	0	0	0	0	1	1	1	2	3	4	7	10
	900	0	0	0	0	1	1	1	2	3	4	6	9
	1000	0	0	0	0	0	1	1	1	3	3	6	8
FEP, FEPB, PFA, PFAH, TFE	14	11	20	33	58	79	131	188	290	389	502	790	1142
	12	8	15	24	42	58	96	137	212	284	366	577	834
	10	6	10	17	30	41	69	98	152	204	263	414	598
	8	3	6	10	17	24	39	56	87	117	150	237	343
	6	2	4	7	12	17	28	40	62	83	107	169	244
	4	1	3	5	8	12	19	28	43	58	75	118	170
	3	1	2	4	7	10	16	23	36	48	62	98	142
	2	1	1	3	6	8	13	19	30	40	51	81	117
PFA, PFAH, TFE	1	1	1	2	4	5	9	13	20	28	36	56	81

TABLE C10 Maximum Number of Conductors and Fixture Wires in Rigid PVC Conduit, Schedule 40 and HDPE Conduit (continued)

Type Letters	*Conductor Size AWG/kcmil*	*Conduit Trade Sizes in Inches*											
		½	*¾*	*1*	*1¼*	*1½*	*2*	*2½*	*3*	*3½*	*4*	*5*	*6*
PFA, PFAH, TFE, Z	1/0	1	1	1	3	4	8	11	17	23	30	47	68
	2/0	0	1	1	3	4	6	9	14	19	24	39	56
	3/0	0	1	1	2	3	5	7	12	16	20	32	46
	4/0	0	1	1	1	2	4	6	9	13	16	26	38
Z	14	13	24	40	70	95	158	226	350	469	605	952	1376
	12	9	17	28	49	68	112	160	248	333	429	675	976
	10	6	10	17	30	41	69	98	152	204	263	414	598
	8	3	6	11	19	26	43	62	96	129	166	261	378
	6	2	4	7	13	18	30	43	67	90	116	184	265
	4	1	3	5	9	12	21	30	46	62	80	126	183
	3	1	2	4	6	9	15	22	34	45	58	92	133
	2	1	1	3	5	7	12	18	28	38	49	77	111
	1	1	1	2	4	6	10	14	23	30	39	62	90
XHH, XHHW, XHHW-2, ZW	14	8	14	24	42	57	94	135	209	280	361	568	822
	12	6	11	18	32	44	72	103	160	215	277	436	631
	10	4	8	13	24	32	54	77	119	160	206	325	470
	8	2	4	7	13	18	30	43	66	89	115	181	261
	6	1	3	5	10	13	22	32	49	66	85	134	193

	4	1	2	4	7	9	16	23	35	48	61	97	140
	3	1	1	3	6	8	13	19	30	40	52	82	118
	2	1	1	3	5	7	11	16	25	34	44	69	99
XHH,	1	1	1	1	3	5	8	12	19	25	32	51	74
XHHW,	1/0	1	1	1	3	4	7	10	16	21	27	43	62
XHHW-2	2/0	0	1	1	2	3	6	8	13	17	23	36	52
	3/0	0	1	1	1	3	5	7	11	14	19	30	43
	4/0	0	1	1	1	2	4	6	9	12	15	24	35
	250	0	0	1	1	1	3	5	7	10	13	20	29
	300	0	0	1	1	1	3	4	6	8	11	17	25
	350	0	0	1	1	1	2	3	5	7	9	15	22
	400	0	0	0	1	1	1	3	5	6	8	13	19
	500	0	0	0	1	1	1	2	4	5	7	11	16
	600	0	0	0	1	1	1	1	3	4	5	9	13
	700	0	0	0	0	1	1	1	3	4	5	8	11
	750	0	0	0	0	1	1	1	2	3	4	7	11
	800	0	0	0	0	1	1	1	2	3	4	7	10
	900	0	0	0	0	1	1	1	2	3	4	6	9
	1000	0	0	0	0	0	1	1	1	3	3	6	8
	1250	0	0	0	0	0	1	1	1	1	3	4	6
	1500	0	0	0	0	0	1	1	1	1	2	4	5
	1750	0	0	0	0	0	0	1	1	1	1	3	5
	2000	0	0	0	0	0	0	1	1	1	1	3	4

TABLE C10 Maximum Number of Conductors and Fixture Wires in Rigid PVC Conduit, Schedule 40 and HDPE Conduit (continued)

Fixture Wires

Type Letters	*Conductor Size AWG/kcmil*	*Conduit Trade Sizes in Inches*					
		½	*¾*	*1*	*1¼*	*1½*	*2*
FFH-2, RFH-2, RFHH-3	18 16	8 6	14 12	23 19	40 33	54 46	90 76
SF-2, SFF-2	18 16 14	10 8 6	17 14 12	29 24 19	50 42 33	69 57 46	114 94 76
SF-1, SFF-1	18	17	31	51	89	122	202
AF, RFH-1, RFHH-2, TF, TFF, XF, XFF	18	13	23	38	66	90	149
AF, RFHH-2, TF, TFF, XF, XFF	16	10	18	30	53	73	120
AF, XF, XFF	14	8	14	24	42	57	94
TFN, TFFN	18 16	20 16	37 28	60 46	105 80	144 110	239 183
PF, PFF, PGF, PGFF, PAF, PTF, PTFF, PAFF	18 16 14	19 15 11	35 27 20	57 44 33	100 77 58	137 106 79	227 175 131

TABLE C10 (continued)

Type Letters	*Conductor Size AWG/kcmil*	*Conduit Trade Sizes in Inches*					
		½	*¾*	*1*	*1¼*	*1½*	*2*
HF, HFF,	18	25	45	74	129	176	292
ZF, ZFF,	16	18	33	54	95	130	216
ZHF	14	13	24	40	70	95	158
KF-2,	18	36	65	107	187	256	424
KFF-2	16	26	46	75	132	180	299
	14	17	31	52	90	124	205
	12	12	22	35	62	85	141
	10	8	14	24	42	57	94
KF-1,	18	43	78	128	223	305	506
KFF-1	16	30	55	90	157	214	355
	14	20	37	60	105	144	239
	12	13	24	40	70	95	158
	10	9	16	26	45	62	103
AF, XF,	12	4	8	12	22	30	50
XFF	10	3	6	10	17	24	39

Note: This table is for concentric stranded conductors only. For compact stranded conductors, Table C10A shall be used.

TABLE C10A Maximum Number of Compact Conductors in Rigid PVC Conduit, Schedule 40 and HDPE Conduit

(Based on Table 1, Chapter 9)

Type Letters	Conductor Size AWG/kcmil	Conduit Trade Sizes in Inches											
		½	¾	1	1¼	1½	2	2½	3	3½	4	5	6
THW, THW-2, THHW	8	1	4	6	11	15	26	37	57	76	98	155	224
	6	1	3	5	9	12	20	28	44	59	76	119	173
	4	1	1	3	6	9	15	21	33	44	57	89	129
	2	1	1	2	5	6	11	15	24	32	42	66	95
	1	1	1	1	3	4	7	11	17	23	29	46	67
	1/0	0	1	1	3	4	6	9	15	20	25	40	58
	2/0	0	1	1	2	3	5	8	12	16	21	34	49
	3/0	0	1	1	1	3	5	7	10	14	18	29	42
	4/0	0	1	1	1	2	4	5	9	12	15	24	35
	250	0	0	1	1	1	3	4	7	9	12	19	27
	300	0	0	1	1	1	2	4	6	8	10	16	24
	350	0	0	1	1	1	2	3	5	7	9	15	21
	400	0	0	0	1	1	1	3	5	6	8	13	19
	500	0	0	0	1	1	1	2	4	5	7	11	16
	600	0	0	0	1	1	1	1	3	4	5	9	13
	700	0	0	0	0	1	1	1	3	4	5	8	12
	750	0	0	0	0	1	1	1	2	3	5	7	11
	1000	0	0	0	0	1	1	1	1	3	4	6	9

THHN,	8	—	—	—	—	—	—	—	—	—	—	—	—
THWN,	6	2	4	7	13	17	29	41	64	86	111	175	253
THWN-2	4	1	2	4	8	11	18	25	40	53	68	108	156
	2	1	1	3	5	8	13	18	28	38	49	77	112
	1	1	1	2	4	6	9	14	21	29	37	58	84
	1/0	1	1	1	3	5	8	12	18	24	31	49	72
	2/0	0	1	1	3	4	7	9	15	20	26	41	59
	3/0	0	1	1	2	3	5	8	12	17	22	34	50
	4/0	0	1	1	1	3	5	6	10	14	18	28	41
	250	0	0	1	1	1	3	5	8	11	14	22	32
	300	0	0	1	1	1	3	4	7	9	12	19	28
	350	0	0	1	1	1	3	4	6	8	10	17	24
	400	0	0	1	1	1	2	3	5	7	9	15	22
	500	0	0	0	1	1	1	3	4	6	8	13	18
	600	0	0	0	1	1	1	2	4	5	6	10	15
	700	0	0	0	1	1	1	1	3	4	5	9	13
	750	0	0	0	1	1	1	1	3	4	5	8	12
	1000	0	0	0	0	1	1	1	2	3	4	6	9

TABLE C10A Maximum Number of Compact Conductors in Rigid PVC Conduit, Schedule 40 and HDPE Conduit (continued)

Type Letters	*Conductor Size AWG/kcmil*	*Conduit Trade Sizes in Inches*											
		½	*¾*	*1*	*1¼*	*1½*	*2*	*2½*	*3*	*3½*	*4*	*5*	*6*
XHHW, XHHW-2	8	3	5	8	14	20	33	47	73	99	127	200	290
	6	1	4	6	11	15	25	35	55	73	94	149	215
	4	1	2	4	8	11	18	25	40	53	68	108	156
	2	1	1	3	5	8	13	18	28	38	49	77	112
	1	1	1	2	4	6	9	14	21	29	37	58	84
	1/0	1	1	1	3	5	8	12	18	24	31	49	72
	2/0	1	1	1	3	4	7	10	15	20	26	42	60
	3/0	0	1	1	2	3	5	8	12	17	22	34	50
	4/0	0	1	1	1	3	5	7	10	14	18	29	42
	250	0	0	1	1	1	4	5	8	11	14	23	33
	300	0	0	1	1	1	3	4	7	9	12	19	28
	350	0	0	1	1	1	3	4	6	8	11	17	25

	400	0	0	1	1	1	2	3	5	7	10	15	22
	500	0	0	0	1	1	1	3	4	6	8	13	18
	600	0	0	0	1	1	1	2	4	5	6	10	15
	700	0	0	0	1	1	1	1	3	4	5	9	13
	750	0	0	0	1	1	1	1	3	4	5	8	12
	1000	0	0	0	0	1	1	1	2	3	4	6	9

Definition: Compact stranding is the result of a manufacturing process where the standard conductor is compressed to the extent that the interstices (voids between strand wires) are virtually eliminated.

TABLE C11 Maximum Number of Conductors and Fixture Wires in Type A, Rigid PVC Conduit
(Based on Table 1, Chapter 9)

Type Letters	*Conductor Size AWG/kcmil*	*Conduit Trade Sizes in Inches*									
		½	*¾*	*1*	*1¼*	*1½*	*2*	*2½*	*3*	*3½*	*4*
RH	14	7	12	20	34	44	70	104	157	204	262
	12	6	10	16	27	35	56	84	126	164	211
RHH, RHW, RHW-2	14	5	9	15	24	31	49	74	112	146	187
	12	4	7	12	20	26	41	61	93	121	155
RH, RHH, RHW, RHW-2	10	3	6	10	16	21	33	50	75	98	125
	8	1	3	5	8	11	17	26	39	51	65
	6	1	2	4	6	9	14	21	31	41	52
	4	1	1	3	5	7	11	16	24	32	41
	3	1	1	3	4	6	9	14	21	28	36
	2	1	1	2	4	5	8	12	18	24	31
	1	0	1	1	2	3	5	8	12	16	20

1/0	0	1	1	2	3	5	7	10	14	18
2/0	0	1	1	1	2	4	6	9	12	15
3/0	0	1	1	1	1	3	5	8	10	13
4/0	0	0	1	1	1	3	4	7	9	11
250	0	0	1	1	1	1	3	5	7	8
300	0	0	1	1	1	1	3	4	6	7
350	0	0	0	1	1	1	2	4	5	7
400	0	0	0	1	1	1	2	4	5	6
500	0	0	0	1	1	1	1	3	4	5
600	0	0	0	0	1	1	1	2	3	4
700	0	0	0	0	1	1	1	2	3	4
750	0	0	0	0	1	1	1	1	3	4
800	0	0	0	0	1	1	1	1	3	3
900	0	0	0	0	0	1	1	1	2	3
1000	0	0	0	0	0	1	1	1	2	3
1250	0	0	0	0	0	1	1	1	1	2
1500	0	0	0	0	0	0	1	1	1	1
1750	0	0	0	0	0	0	1	1	1	1
2000	0	0	0	0	0	0	1	1	1	1

TABLE C11 Maximum Number of Conductors and Fixture Wires in Type A, Rigid PVC Conduit (continued)

Type Letters	*Conductor Size AWG/kcmil*	*Conduit Trade Sizes in Inches*									
		½	*¾*	*1*	*1¼*	*1½*	*2*	*2½*	*3*	*3½*	*4*
TW	14	11	18	31	51	67	105	157	235	307	395
	12	8	14	24	39	51	80	120	181	236	303
	10	6	10	18	29	38	60	89	135	176	226
	8	3	6	10	16	21	33	50	75	98	125
RHH*, RHW*, RHW-2*, THHW, THW, THW-2	14	7	12	20	34	44	70	104	157	204	262
RHH*, RHW*, RHW-2*, THHW, THW	12	6	10	16	27	35	56	84	126	164	211
	10	4	8	13	21	28	44	65	98	128	165
RHH*, RHW*, RHW-2*, THHW, THW, THW-2	8	2	4	8	12	16	26	39	59	77	98

RHH*, RHW*,	6	1	3	6	9	13	20	30	45	59	75
RHW-2*,	4	1	2	4	7	9	15	22	33	44	56
TW, THW,	3	1	1	4	6	8	13	19	29	37	48
THHW,	2	1	1	3	5	7	11	16	24	32	41
THW-2	1	1	1	1	3	5	7	11	17	22	29
	1/0	1	1	1	3	4	6	10	14	19	24
	2/0	0	1	1	2	3	5	8	12	16	21
	3/0	0	1	1	1	3	4	7	10	13	17
	4/0	0	1	1	1	2	4	6	9	11	14
	250	0	0	1	1	1	3	4	7	9	12
	300	0	0	1	1	1	2	4	6	8	10
	350	0	0	1	1	1	2	3	5	7	9
	400	0	0	1	1	1	1	3	5	6	8
	500	0	0	0	1	1	1	2	4	5	7
	600	0	0	0	1	1	1	1	3	4	5
	700	0	0	0	1	1	1	1	3	4	5
	750	0	0	0	1	1	1	1	3	3	4
	800	0	0	0	0	1	1	1	2	3	4
	900	0	0	0	0	1	1	1	2	3	4
	1000	0	0	0	0	1	1	1	1	3	3
	1250	0	0	0	0	0	1	1	1	1	3
	1500	0	0	0	0	0	1	1	1	1	2
	1750	0	0	0	0	0	0	1	1	1	1
	2000	0	0	0	0	0	0	1	1	1	1

* Types RHH, RHW, and RHW-2 without outer covering.

TABLE C11 Maximum Number of Conductors and Fixture Wires in Type A, Rigid PVC Conduit (continued)

Type Letters	*Conductor Size AWG/kcmil*	*Conduit Trade Sizes in Inches*									
		½	*¾*	*1*	*1¼*	*1½*	*2*	*2½*	*3*	*3½*	*4*
THHN, THWN, THWN-2	14	16	27	44	73	96	150	225	338	441	566
	12	11	19	32	53	70	109	164	246	321	412
	10	7	12	20	33	44	69	103	155	202	260
	8	4	7	12	19	25	40	59	89	117	150
	6	3	5	8	14	18	28	43	64	84	108
	4	1	3	5	8	11	17	26	39	52	66
	3	1	2	4	7	9	15	22	33	44	56
	2	1	1	3	6	8	12	19	28	37	47
	1	1	1	2	4	6	9	14	21	27	35
	1/0	1	1	2	4	5	8	11	17	23	29
	2/0	1	1	1	3	4	6	10	14	19	24
	3/0	0	1	1	2	3	5	8	12	16	20
	4/0	0	1	1	1	3	4	6	10	13	17
	250	0	1	1	1	2	3	5	8	10	14
	300	0	0	1	1	1	3	4	7	9	12
	350	0	0	1	1	1	2	4	6	8	10
	400	0	0	1	1	1	2	3	5	7	9

	500	0	0	1	1	1	1	3	4	6	7
	600	0	0	0	1	1	1	2	3	5	6
	700	0	0	0	1	1	1	1	3	4	5
	750	0	0	0	1	1	1	1	3	4	5
	800	0	0	0	1	1	1	1	3	4	5
	900	0	0	0	0	1	1	1	2	3	4
	1000	0	0	0	0	1	1	1	2	3	4
FEP, FEPB,	14	15	26	43	70	93	146	218	327	427	549
PFA, PFAH,	12	11	19	31	51	68	106	159	239	312	400
TFE	10	8	13	22	37	48	76	114	171	224	287
	8	4	8	13	21	28	44	65	98	128	165
	6	3	5	9	15	20	31	46	70	91	117
	4	1	4	6	10	14	21	32	49	64	82
	3	1	3	5	8	11	18	27	40	53	68
	2	1	2	4	7	9	15	22	33	44	56
PFA, PFAH, TFE	1	1	1	3	5	6	10	15	23	30	39

TABLE C11 Maximum Number of Conductors and Fixture Wires in Type A, Rigid PVC Conduit (continued)

Type Letters	*Conductor Size AWG/kcmil*	*Conduit Trade Sizes in Inches*									
		½	*¾*	*1*	*1¼*	*1½*	*2*	*2½*	*3*	*3½*	*4*
PFA, PFAH, TFE, Z	1/0	1	1	2	4	5	8	13	19	25	32
	2/0	1	1	1	3	4	7	10	16	21	27
	3/0	1	1	1	3	3	6	9	13	17	22
	4/0	0	1	1	2	3	5	7	11	14	18
Z	14	18	31	52	85	112	175	263	395	515	661
	12	13	22	37	60	79	124	186	280	365	469
	10	8	13	22	37	48	76	114	171	224	287
	8	5	8	14	23	30	48	72	108	141	181
	6	3	6	10	16	21	34	50	76	99	127
	4	2	4	7	11	15	23	35	52	68	88
	3	1	3	5	8	11	17	25	38	50	64
	2	1	2	4	7	9	14	21	32	41	53
	1	1	1	3	5	7	11	17	26	33	43
XHH, XHHW, XHHW-2, ZW	14	11	18	31	51	67	105	157	235	307	395
	12	8	14	24	39	51	80	120	181	236	303
	10	6	10	18	29	38	60	89	135	176	226
	8	3	6	10	16	21	33	50	75	98	125
	6	2	4	7	12	15	24	37	55	72	93

	4	1	3	5	8	11	18	26	40	52	67
	3	1	2	4	7	9	15	22	34	44	57
	2	1	1	3	6	8	12	19	28	37	48
XHH, XHHW, XHHW-2	1	1	1	3	4	6	9	14	21	28	35
	1/0	1	1	2	4	5	8	12	18	23	30
	2/0	1	1	1	3	4	6	10	15	19	25
	3/0	0	1	1	2	3	5	8	12	16	20
	4/0	0	1	1	1	3	4	7	10	13	17
	250	0	1	1	1	2	3	5	8	11	14
	300	0	0	1	1	1	3	5	7	9	12
	350	0	0	1	1	1	3	4	6	8	10
	400	0	0	1	1	1	2	3	5	7	9
	500	0	0	1	1	1	1	3	4	6	8
	600	0	0	0	1	1	1	2	3	5	6
	700	0	0	0	1	1	1	1	3	4	5
	750	0	0	0	1	1	1	1	3	4	5
	800	0	0	0	1	1	1	1	3	4	5
	900	0	0	0	0	1	1	1	2	3	4
	1000	0	0	0	0	1	1	1	2	3	4
	1250	0	0	0	0	0	1	1	1	2	3
	1500	0	0	0	0	0	1	1	1	1	2
	1750	0	0	0	0	0	1	1	1	1	2
	2000	0	0	0	0	0	0	1	1	1	1

TABLE C11 Maximum Number of Conductors and Fixture Wires in Type A, Rigid PVC Conduit (continued)

Fixture Wires

Type Letters	*Conductor Size AWG/kcmil*	*Conduit Trade Sizes in Inches*					
		½	*¾*	*1*	*1¼*	*1½*	*2*
FFH-2, RFH-2, RFHH-3 RFHH-3	18 16	10 9	18 15	30 25	48 41	64 54	100 85
SF-2, SFF-2	18 16 14	13 11 9	22 18 15	37 31 25	61 51 41	81 67 54	127 105 85
SF-1, SFF-1	18	23	40	66	108	143	224
AF, RFH-1, RFHH-2, TF, TFF, XF, XFF	18	17	29	49	80	105	165
AF, RFHH-2, TF, TFF, XF, XFF	16	14	24	39	65	85	134
AF, XF, XFF	14	11	18	31	51	67	105
TFN, TFFN	18 16	28 21	47 36	79 60	128 98	169 129	265 202
PF, PFF, PGF, PGFF, PAF, PTF, PTFF, PAFF	18 16 14	26 20 15	45 34 26	74 58 43	122 94 70	160 124 93	251 194 146

TABLE C11 (continued)

Type Letters	*Conductor Size AWG/kcmil*	*Conduit Trade Sizes in Inches*					
		½	*¾*	*1*	*1¼*	*1½*	*2*
HF, HFF, ZF, ZFF, ZHF	18	34	58	96	157	206	324
	16	25	42	71	116	152	239
	14	18	31	52	85	112	175
KF-2, KFF-2	18	49	84	140	228	300	470
	16	35	59	98	160	211	331
	14	24	40	67	110	145	228
	12	16	28	46	76	100	157
	10	11	18	31	51	67	105
KF-1, KFF-1	18	59	100	167	272	357	561
	16	41	70	117	191	251	394
	14	28	47	79	128	169	265
	12	18	31	52	85	112	175
	10	12	20	34	55	73	115
AF, XF, XFF	12	6	10	16	27	35	56
	10	4	8	13	21	28	44

Note: This table is for concentric stranded conductors only. For compact stranded conductors, Table C11A shall be used.

TABLE C11A Maximum Number of Compact Conductors in Type A, Rigid PVC Conduit

(Based on Table 1, Chapter 9)

Type Letters	Conductor Size AWG/kcmil	Conduit Trade Sizes in Inches									
		½	¾	1	1¼	1½	2	2½	3	3½	4
THW, THW-2, THHW	8	3	5	8	14	18	28	42	64	84	107
	6	2	4	6	10	14	22	33	49	65	83
	4	1	3	5	8	10	16	24	37	48	62
	2	1	1	3	6	7	12	18	27	36	46
	1	1	1	2	4	5	8	13	19	25	32
	1/0	1	1	1	3	4	7	11	16	21	28
	2/0	1	1	1	3	4	6	9	14	18	23
	3/0	0	1	1	2	3	5	8	12	15	20
	4/0	0	1	1	1	3	4	6	10	13	17
	250	0	1	1	1	1	3	5	8	10	13
	300	0	0	1	1	1	3	4	7	9	11
	350	0	0	1	1	1	2	4	6	8	10
	400	0	0	1	1	1	2	3	5	7	9
	500	0	0	1	1	1	1	3	4	6	8
	600	0	0	0	1	1	1	2	3	5	6
	700	0	0	0	1	1	1	1	3	4	5
	750	0	0	0	1	1	1	1	3	4	5
	1000	0	0	0	0	1	1	1	2	3	4

THHN, THWN, THWN-2	8	—	—	—	—	—	—	—	—	—	—
	6	3	5	9	15	20	32	48	72	94	121
	4	1	3	6	9	12	20	30	45	58	75
	2	1	2	4	7	9	14	21	32	42	54
	1	1	1	3	5	7	10	16	24	31	40
	1/0	1	1	2	4	6	9	13	20	27	34
	2/0	1	1	1	3	5	7	11	17	22	28
	3/0	1	1	1	3	4	6	9	14	18	24
	4/0	0	1	1	2	3	5	8	11	15	19
	250	0	1	1	1	2	4	6	9	12	15
	300	0	1	1	1	1	3	5	8	10	13
	350	0	0	1	1	1	3	4	7	9	11
	400	0	0	1	1	1	2	4	6	8	10
	500	0	0	1	1	1	2	3	5	7	9
	600	0	0	0	1	1	1	3	4	5	7
	700	0	0	0	1	1	1	2	3	5	6
	750	0	0	0	1	1	1	2	3	4	6
	1000	0	0	0	0	1	1	1	2	3	4

TABLE C11A Maximum Number of Compact Conductors in Type A, Rigid PVC Conduit (continued)

Type Letters	Conductor Size AWG/kcmil	Conduit Trade Sizes in Inches									
		1½	¾	1	1¼	1½	2	2½	3	3½	4
XHHW,	8	4	6	11	18	23	37	55	83	108	139
XHHW-2	6	3	5	8	13	17	27	41	62	80	103
	4	1	3	6	9	12	20	30	45	58	75
	2	1	2	4	7	9	14	21	32	42	54
	1	1	1	3	5	7	10	16	24	31	40
	1/0	1	1	2	4	6	9	13	20	27	34
	2/0	1	1	1	3	5	7	11	17	22	29
	3/0	1	1	1	3	4	6	9	14	18	24
	4/0	0	1	1	2	3	5	8	12	15	20
	250	0	1	1	1	2	4	6	9	12	16
	300	0	1	1	1	1	3	5	8	10	13
	350	0	0	1	1	1	3	5	7	9	12
	400	0	0	1	1	1	3	4	6	8	11
	500	0	0	1	1	1	2	3	5	7	9
	600	0	0	0	1	1	1	3	4	5	7
	700	0	0	0	1	1	1	2	3	5	6
	750	0	0	0	1	1	1	2	3	4	6
	1000	0	0	0	0	1	1	1	2	3	4

Definition: Compact stranding is the result of a manufacturing process where the standard conductor is compressed to the extent that the interstices (voids between strand wires) are virtually eliminated.

TABLE C12 Maximum Number of Conductors in Type EB, PVC Conduit

(Based on Table 1, Chapter 9)

Type Letters	*Conductor Size AWG/kcmil*	*Conduit Trade Sizes in Inches*					
		2	*3*	*3½*	*4*	*5*	*6*
RH	14	74	166	217	276	424	603
	12	59	134	175	222	341	485
RHH, RHW, RHW-2	14	53	119	155	197	303	430
	12	44	98	128	163	251	357
RH, RHH, RHW, RHW-2	10	35	79	104	132	203	288
	8	18	41	54	69	106	151
	6	15	33	43	55	85	121
	4	11	26	34	43	66	94
	3	10	23	30	38	58	83
	2	9	20	26	33	50	72
	1	6	13	17	21	33	47
	1/0	5	11	15	19	29	41
	2/0	4	10	13	16	25	36
	3/0	4	8	11	14	22	31
	4/0	3	7	9	12	18	26
	250	2	5	7	9	14	20
	300	1	5	6	8	12	17
	350	1	4	5	7	11	16
	400	1	4	5	6	10	14
	500	1	3	4	5	9	12
	600	1	3	3	4	7	10
	700	1	2	3	4	6	9
	750	1	2	3	4	6	9
	800	1	2	3	4	6	8
	900	1	1	2	3	5	7
	1000	1	1	2	3	5	7
	1250	1	1	1	2	3	5
	1500	0	1	1	1	3	4
	1750	0	1	1	1	3	4
	2000	0	1	1	1	2	3

TABLE C12 Maximum Number of Conductors in Type EB, PVC Conduit (continued)

Type Letters	*Conductor Size AWG/kcmil*	*Conduit Trade Sizes in Inches*					
		2	*3*	*3½*	*4*	*5*	*6*
TW	14	111	250	327	415	638	907
	12	85	192	251	319	490	696
	10	63	143	187	238	365	519
	8	35	79	104	132	203	288
RHH*, RHW*, RHW-2*, THHW, THW, THW-2	14	74	166	217	276	424	603
RHH*, RHW*, RHW-2*, THHW, THW	12	59	134	175	222	341	485
	10	46	104	136	173	266	378
RHH*, RHW*, RHW-2*, THHW, THW, THW-2	8	28	62	81	104	159	227
RHH*, RHW*, RHW-2*, TW, THW, THHW, THW-2	6	21	48	62	79	122	173
	4	16	36	46	59	91	129
	3	13	30	40	51	78	111
	2	11	26	34	43	66	94
	1	8	18	24	30	46	66
	1/0	7	15	20	26	40	56
	2/0	6	13	17	22	34	48
	3/0	5	11	14	18	28	40
	4/0	4	9	12	15	24	34
	250	3	7	10	12	19	27
	300	3	6	8	11	17	24
	350	2	6	7	9	15	21
	400	2	5	7	8	13	19
	500	1	4	5	7	11	16
	600	1	3	4	6	9	13
	700	1	3	4	5	8	11
	750	1	3	4	5	7	11
	800	1	3	3	4	7	10
	900	1	2	3	4	6	9
	1000	1	2	3	4	6	8
	1250	1	1	2	3	4	6
	1500	1	1	1	2	4	6
	1750	1	1	1	2	3	5
	2000	0	1	1	1	3	4

* Types RHH, RHW, and RHW-2 without outer covering.

TABLE C12 Maximum Number of Conductors in Type EB, PVC Conduit (continued)

Type Letters	Conductor Size AWG/kcmil	Conduit Trade Sizes in Inches					
		2	3	3½	4	5	6
THHN, THWN, THWN-2	14	159	359	468	595	915	1300
	12	116	262	342	434	667	948
	10	73	165	215	274	420	597
	8	42	95	124	158	242	344
	6	30	68	89	114	175	248
	4	19	42	55	70	107	153
	3	16	36	46	59	91	129
	2	13	30	39	50	76	109
	1	10	22	29	37	57	80
	1/0	8	18	24	31	48	68
	2/0	7	15	20	26	40	56
	3/0	5	13	17	21	33	47
	4/0	4	10	14	18	27	39
	250	4	8	11	14	22	31
	300	3	7	10	12	19	27
	350	3	6	8	11	17	24
	400	2	6	7	10	15	21
	500	1	5	6	8	12	18
	600	1	4	5	6	10	14
	700	1	3	4	6	9	12
	750	1	3	4	5	8	12
	800	1	3	4	5	8	11
	900	1	3	3	4	7	10
	1000	1	2	3	4	6	9
FEP, FEPB, PFA, PFAH, TFE	14	155	348	454	578	888	1261
	12	113	254	332	422	648	920
	10	81	182	238	302	465	660
	8	46	104	136	173	266	378
	6	33	74	97	123	189	269
	4	23	52	68	86	132	188
	3	19	43	56	72	110	157
	2	16	36	46	59	91	129
PFA, PFAH, TFE	1	11	25	32	41	63	90

TABLE C12 Maximum Number of Conductors in Type EB, PVC Conduit (continued)

Type Letters	Conductor Size AWG/kcmil	Conduit Trade Sizes in Inches					
		2	3	3½	4	5	6
PFA, PFAH, TFE, Z	1/0	9	20	27	34	53	75
	2/0	7	17	22	28	43	62
	3/0	6	14	18	23	36	51
	4/0	5	11	15	19	29	42
Z	14	186	419	547	696	1069	1519
	12	132	297	388	494	759	1078
	10	81	182	238	302	465	660
	8	51	115	150	191	294	417
	6	36	81	105	134	206	293
	4	24	55	72	92	142	201
	3	18	40	53	67	104	147
	2	15	34	44	56	86	122
	1	12	27	36	45	70	99
XHH, XHHW, XHHW-2, ZW	14	111	250	327	415	638	907
	12	85	192	251	319	490	696
	10	63	143	187	238	365	519
	8	35	79	104	132	203	288
	6	26	59	77	98	150	213
	4	19	42	56	71	109	155
	3	16	36	47	60	92	131
	2	13	30	39	50	77	110
XHH, XHHW, XHHW-2	1	10	22	29	37	58	82
	1/0	8	19	25	31	48	69
	2/0	7	16	20	26	40	57
	3/0	6	13	17	22	33	47
	4/0	5	11	14	18	27	39
	250	4	9	11	15	22	32
	300	3	7	10	12	19	28
	350	3	6	9	11	17	24
	400	2	6	8	10	15	22
	500	1	5	6	8	12	18
	600	1	4	5	6	10	14
	700	1	3	4	6	9	12
	750	1	3	4	5	8	12
	800	1	3	4	5	8	11
	900	1	3	3	4	7	10
	1000	1	2	3	4	6	9
	1250	1	1	2	3	5	7
	1500	1	1	1	3	4	6
	1750	1	1	1	2	4	5
	2000	0	1	1	1	3	5

Note: This table is for concentric stranded conductors only. For compact stranded conductors, Table 12A should be used.

TABLE C12A Maximum Number of Compact Conductors in Type EB, PVC Conduit

(Based on Table 1, Chapter 9)

Type Letters	Conductor Size AWG/kcmil	Conduit Trade Sizes in Inches					
		2	3	3½	4	5	6
THW, THW-2, THHW	8	30	68	89	113	174	247
	6	23	52	69	87	134	191
	4	17	39	51	65	100	143
	2	13	29	38	48	74	105
	1	9	20	26	34	52	74
	1/0	8	17	23	29	45	64
	2/0	6	15	19	24	38	54
	3/0	5	12	16	21	32	46
	4/0	4	10	14	17	27	38
	250	3	8	11	14	21	30
	300	3	7	9	12	19	26
	350	3	6	8	11	17	24
	400	2	6	7	10	15	21
	500	1	5	6	8	12	18
	600	1	4	5	6	10	14
	700	1	3	4	6	9	13
	750	1	3	4	5	8	12
	1000	1	2	3	4	7	9
THHN, THWN, THWN-2	8	-	-	-	-	-	-
	6	34	77	100	128	196	279
	4	21	47	62	79	121	172
	2	15	34	44	57	87	124
	1	11	25	33	42	65	93
	1/0	9	22	28	36	56	79
	2/0	8	18	23	30	46	65
	3/0	6	15	20	25	38	55
	4/0	5	12	16	20	32	45
	250	4	10	13	16	25	35
	300	4	8	11	14	22	31
	350	3	7	9	12	19	27
	400	3	6	8	11	17	24
	500	2	5	7	9	14	20
	600	1	4	6	7	11	16
	700	1	4	5	6	10	14
	750	1	4	5	6	9	14
	1000	1	3	3	4	7	10

TABLE C12A Maximum Number of Compact Conductors in Type EB, PVC Conduit (continued)

Type Letters	*Conductor Size AWG/kcmil*	*Conduit Trade Sizes in Inches*					
		2	*3*	*3½*	*4*	*5*	*6*
XHHW, XHHW-2	8	39	88	115	146	225	320
	6	29	65	85	109	167	238
	4	21	47	62	79	121	172
	2	15	34	44	57	87	124
	1	11	25	33	42	65	93
	1/0	9	22	28	36	56	79
	2/0	8	18	24	30	47	67
	3/0	6	15	20	25	38	55
	4/0	5	12	16	21	32	46
	250	4	10	13	17	26	37
	300	4	8	11	14	22	31
	350	3	7	10	12	19	28
	400	3	7	9	11	17	25
	500	2	5	7	9	14	20
	600	1	4	6	7	11	16
	700	1	4	5	6	10	14
	750	1	3	5	6	9	13
	1000	1	3	4	5	7	10

Definition: Compact stranding is the result of a manufacturing process where the standard conductor is compressed to the extent that the interstices (voids between strand wires) are virtually eliminated.